INTRODUCTION TO AGRICULTURAL BOTANY

Others Related Books

INTRODUCTION TO AGRICULTURAL BOTANY

Dr. Muthaiyan Pandiyan
Ph.D., FNABS and FJPN, DEAN,
Agricultural College and Research Institute,
Tamil Nadu Agriculture University (T.N.A.U.),
Vazhavachanur, Tiruvannamalai,
Tamil Nadu, INDIA

SCIENTIFIC PUBLISHERS

Published by
SCIENTIFIC PUBLISHERS (INDIA) 5 A, New Pali Road, P.O. Box 91 Jodhpur 342 001 (INDIA)
E-mail: info@scientificpub.com
Website: http://www.scientificpubonline.com

Print 2026

ISBN: 978-93-89832-80-8 (PB)
ISBN: 978-93-89832-81-5 (EBook)

Printed in India

PREFACE

The book "**Introduction to Agricultural Botany**" has been written covering 21 families of **Field and Horticultural crops**, for the students of all agricultural universities of India. The undergraduate & post-graduate students of Botany subject of conventional universities of the country will also be benefited with this new type of book. The book covers nearly 72 crops, in 20 chapters where 3 chapters with fundamentals of botany and 13 chapters with botanical description of nearly 72 field and horticultural crops. Summary of botanical description of important cultivated crops is given in one chapter and another 3 chapter containg model questions. Each chapter of model question consists of midsemester, final theory and final practical examination questions.

The book has been written in simple English & short format with suitable diagrams. This will be useful for student to easily understand the subject at both undergraduate and post-graduate level. Additionally, this book also has question bank for five respective units and also model question for mid semester, final theory and practical examinations.

Hope this book would be helpful for undergraduates and post-graduates students of agriculture. Additionally, question bank for each units will provide strong knowledge to students for preparing higher studies as well as competitive exams of DBT-JRF, ICAR-NET and ARS exams.

Dr. Muthiyan Pandiyan

PBG 101 INTRODUCTIONS TO AGRICULTURAL BOTANY (1+1)

Aim: To expose the students to the basic features of botanical description, economic parts and economic importance of different field and horticultural crops

SYLLABUS FOR THEORY

Unit I: Systems of classification and general morphological description

Bentham and Hooker's classification of plant kingdom — International code of nomenclature and its major guidelines – author citation – Agricultural classification of crops; General morphology: Life span, habit, root, stem, leaf - petiole, leaf margin, leaf apex, leaf shape, venation and phyllotaxy; Modification of roots and leaf; Floral morphology: Kinds of bracts, inflorescence; Structure of flower, androecium, gynoecium, placentation, types of fruits.

Unit II: Botanical description and economic uses of Poaceae

List of cultivated crops, economic parts, chromosome number and family description of Poaceae: Key botanical features of Rice, Wheat, Sorghum, Maize, Pearl millet, Finger millet, list of small millets, Guinea grass, Napier grass, *Cenchrus* and Sugarcane

Unit III: Botanical description and economic uses of Papilionaceae

List of cultivated crops, economic parts, chromosome number and family description of Papilionaceae: Key botanical features of Red gram, Bengal gram, Soybean, Black gram, Green gram, Cowpea, Lablab, Horse gram, Groundnut, Lucerne, *Stylosanthes*, Clitoria, Agathi and Sunnhemp,

Unit IV: Botanical description and economic uses of Pedaliaceae, Asteraceae, Oleaceae, Brassicaceae, Euphorbiaceae, Arecaceae and Malvaceae

List of cultivated crops, economic parts, chromosome number and family description of the following families and Key botanical features of the crops given against them: Pedaliaceae - Gingelly; Asteraceae - Sunflower, Safflower, Chrysanthemum; Oleaceae – Jasmine; Brassicaceae - Rapeseed and Mustard, Cabbage, Cauliflower;

Euphorbiaceae: Castor; Jatropha and Tapioca; Arecaceae: Coconut, Arecanut, Oilpalm, Sugarpalm; Malvaceae: Cotton, Mesta and Bhendi.

Unit V: Botanical description and economic uses of Tiliaceae, Piperaceae, Chenopodiaceae, Solanaceae, Mimosae, Moraceae, Cucurbitaceae, Alliaceae, Musaceae, Rubiaceae, Theaceae

List of cultivated crops, economic parts, chromosome number and family description of the following families and key botanical features of the crops given against them. Tiliaceae: Jute; Piperaceae: Betelvine; Chenopodiaceae: Sugar beet; Solanaceae: Tobacco, Potato, Chilli, Tomato and Brinjal; Mimosae: Desmanthes, Subabul and Acacia; Moraceae: Mulberry; Cucurbitaceae: Cucumber, Pumpkin, Ashgourd; Alliaceae: Onion and Garlic; Musaceae: Banana, Manila hemp; Rubiaceae: Coffee; Theaceae: Tea

SYLLABUS FOR PRACTICAL

Family features - observation and description of habit, morphology of root, stem, leaves, inflorescence, flowers, floral diagram, floral formula and economic parts of Poaceae: Rice, Wheat, Sorghum, Maize, Pearl millet, Finger millet, Guinea grass, Napier grass, *Cenchrus* and Sugarcane; Papilionaceae: Redgram, Bengal gram, Soybean, Blackgram, Greengram, Cowpea, Lab-lab, Horse gram, Groundnut, Lucerne, *Stylosanthes*, Clitoria, Agathi and Sunnhemp; Pedaliaceae: Gingelly; Asteraceae: Sunflower, Safflower and Chrysanthemum; Oleaceae: Jasmine; Brassicaceae: Rape and Mustard, Cabbage, Cauliflower; Euphorbiaceae: Castor, Jatropha, Tapioca; Arecaceae: Coconut, Arecanut, Oilpalm and Sugar palm; Malvaceae: Cotton, Mesta, Bhendi; Tiliaceae: Jute; Piperaceae: Betelvine; Chenopodiaceae: Sugar beet; Solanaceae: Tobacco, Potato, Chilli, Tomato and Brinjal; Mimosae: Desmanthes, Subabul and Acacia; Moraceae: Mulberry; Cucurbitaceae: Cucumber, Pumpkin, Ashgourd; Alliaceae: Onion and Garlic; Musaceae: Banana, Manila hemp; Rubiaceae: Coffee: Theaceae: Tea

Theory schedule (No of classes to be covered by the teacher is 17 which include midsemester)

1. Bentham and Hooker's classification of plant kingdom — International code of nomenclature and its major guidelines – author citation – Agricultural classification of crops
2. General morphology: Life span, habit, root, stem, leaf - petiole, leaf margin, leaf apex, leaf shape, venation and phyllotaxy; Modification of roots and leaf
3. Floral morphology: Kinds of bracts, inflorescence; Structure of flower, androecium, gynoecium, placentation , types of fruits.
4. List of cultivated crops, economic parts, chromosome number and family description of Poaceae; Key botanical features of Rice and Wheat.
5. Key botanical features of sorghum, maize, pearl millet and finger millet. List of small millets
6. Key botanical features of Guinea grass, Napier grass, Cenchrus and sugarcane.
7. List of cultivated crops, economic parts, chromosome number and family description of (Papilionaceae) Key botanical features of Red gram, Bengal gram and Soybean
8. Key botanical features of Black gram, Green gram, Cowpea, Lab lab, Horse gram and Groundnut.
9. Mid Semester Examination
10. Key botanical features of Lucerne, Stylosanthes, Clitoria, Agathi, and Sunnhemp.
11. List of cultivated crops, economic parts, chromosome number and family description of Pedaliaceae and Asteraceae: Key botanical features of Gingelly, Sunflower, Safflower, Chrysanthemum; Oleaceae: Jasmine

12. List of cultivated crops, economic parts, chromosome number and family description of Brassicaceae and Euphorbiaceae; Key botanical features of Rapeseed and Mustard, Cabbage, Cauliflower, Castor, Jatropha and Tapioca
13. List of cultivated crops, economic parts, chromosome number and family description of Arecaceae and Malvaceae; Key botanical features of Coconut, Arecanut, Oilpalm, Sugarpalm, Cotton, Mesta and Bhendi.
14. List of cultivated crops, economic parts, chromosome number and family description of Tiliaceae, Piperaceae and Chenopodiaceae; Key botanical features of Jute, Betelvine, Sugar beet.
15. List of cultivated crops, economic parts, chromosome number and family description of Solanaceae, Mimosae and Moraceae; Key botanical features of Tobacco, Potato, Chilli, Tomato and Brinjal, Desmanthes. Subabul, Mulberry
16. List of cultivated crops, economic parts, chromosome number and family description of Cucurbitaceae and Alliaceae; Cucurbitaceae: Key botanical features of Cucumber, Pumpkin, Ashgourd; Alliaceae: Onion and Garlic
17. List of cultivated crops, economic parts, chromosome number and family description of Musaceae, Rubiaceae and Theaceae; Key botanical features of Banana, Manila hemp,Coffee and Tea

Final Theory Examination
Practical schedule

1. Observing general morphology of roots, stems and leaves.
2. Observing general morphology of inflorescence - flowers, stamens and pistils.
3. Family characters, Botany, Economic parts, Floral diagram and Floral formula of the following crop plants:- Poaceae: Rice and Wheat
4. Poaceae: Sorghum, Maize, Pearl millet, Finger millet.
5. Poaceae: Guinea grass, Napier grass, Cenchrus and Sugarcane.
6. Papilionaceae: Redgram, Bengal gram and Soybean.
7. Papilionaceae: Blackgram, Greengram, Cowpea, Lab-lab, Horse gram and Groundnut.
8. Papilionaceae: Lucerne, Stylosanthes, Clitoria, Agathi, Sunnhemp, and Sesbania.
9. Pedaliaceae: Gingelly; Asteraceae: Sunflower, Safflower and Chrysanthemum; Oleaceae: Jasmine
10. Brassicaceae: Rapeseed and Mustard, Cabbage, Cauliflower.
11. Euphorbiaceae: Castor, Jatropha, Tapioca; Arecaceae: Coconut, Arecanut, Oilpalm and Sugar palm.
12. Malvaceae: Cotton, Mesta, Bhendi
13. Tiliaceae: Jute; Piperaceae: Betelvine; Chenopodiaceae: Sugar beet;
14. Solanaceae: Tobacco, Potato, Chilli, Tomato and Brinjal; Mimosae: Desmanthes, Subabul , Moraceae:Mulberry
15. Cucurbitaceae: Cucumber, Pumpkin, Ashgourd; Alliaceae: Onion and Garlic
16. Musaceae: Banana, Manila hemp; Rubiaceae: Coffee; Theaceae: Tea
17. Final Practical Examination

Assignment

- Collection and preparation of 25 herbarium specimens representing minimum of ten families of the crop species studied.

Outcome

- Botanical features and economic importance of different crop plants belonging to 20 families will be exposed.

CONTENTS

Chapter

1

Bentham and Hookers' Classification of Plant Kingdom

INTRODUCTION

Taxonomy: Taxis + nomos

It includes two Greek words

Taxis – arrangement, Nomos – laws

Taxonomy concerns with laws governing classification of plants. The word 'taxonomy' was coined by A.P. de Candole. Taxonomy may deal with plants or animal. Plant taxonomy is called **systematic botany**.

BASIS OF PLANT TAXONOMY

1. Identification
2. Nomenclature
3. Classification

1. **Identification:** It is necessary to assess whether a collected plant is entirely new or already knows. This can be done with the help of herbarium / flora.
    - **Herbarium:** It is a collection of dried and preserved plant specimen which are arranged according to system of classification.
    - **Flora:** It is a book containing the details of habitat and distribution of plant species of particular area
2. **Nomenclature:** Giving a correct scientific name to an identified plant.

    Binomial Nomenclature:

    - It is giving a correct scientific name to an identified organism with two Latin wards.
    - First word is Generic name and it starts with capital letter.
    - Second word is species name and it starts with small letter.
    - Scientific names should be italics or underlined separately.
    - **AIM:** To avoid confusion which comes due to different common names for same plant.
    - **Earlier, plant names were polynomial:**

      Eg. *Caryophyllum saxatilis folis gramineus umbellatis corymbis*

Which means Caryophyllum growing on rocks having grass like leaves and umbellate corymbose inflorescence. This type of polynomial nomenclature is not valid now. The valid nomenclature is binomial nomenclature.

- Binomial nomenclature was first introduced by Gaspard Bauhin but it was strictly followed by Carl Linnaeus in his book 'Species Plantarum'.

ICBN: International Code of Botanical Nomenclature. ICBN frames some rules for nomenclature

Rules of ICBN:

- Every plant should have only one valid scientific name all over the world.
- Binomial nomenclature should be strictly followed.
- Scientific name of plant should be in two Latin words.
- Scientific name of plant should be in italics or should be underlined separately.
- **Author citation:** The person who publishes the description of any plant species or giving a new name to plant is considered as author. The name of author should be given in abbreviated form at the end of scientific name is called author citation.

  Eg. *Oryza sativa* (L.) – Linnaeus described it

  Pennisetum glaucum R. Br. – Robert Brown described it
- **Type specimen:** While naming a new plant, the plant specimen should be prepared with its original description mounted on herbarium sheets and the same should be preserved in any of the recognized Herbarium. It is called 'type specimen'.

Recognized Herbaria

- Royal Botanical Garden, Kew, London
- Indian Botanical Garden, Kolkata, India
- Botanical Survey of India, Coimbatore, Tamil Nadu

- **Tautonyms:** Giving the same ward to Generic and species name. It is not accepted in the system of classification. Eg. *Malua malus*

3. **Classification:** Grouping of living organism systematically called biosystematics. There are three systems of classification.
   - Artificial system
   - Natural system
   - Phylogenetic system (Cladistics)

   (i) Artificial system: This system of classification is based on one or two easily recognized characters.

   Eg. Theophrastus classified plants into Trees, Shrubs, Under shrubs and Herbs. He is called 'Father of Botany'.

   (ii) Natural system: This system of classification is based on natural similarities of vegetative and floral characters. Eg. Bentham and Hooker classified plants into Cryptogams and Phanerogams.

(iii) Phylogenetic system / Cladistics: This system of classification is based evolutionary and genetic relationship of an organism. Eg. Engler and Prantle classified Bacteria and plants into 14 divisions.

Units of classification: The classification system is made up of different units called 'taxon' and they are arranged in hierarchical sequence.

Kingdom
Phyllum / Division
Class
Order
Family
Genus
Species

The basic unit of classification is **'Species'** which is defined as group of individuals which resemble in their morphological and reproductive characters and interbreed among themselves to produce fertile offsprings.

Systematic Position of Rice

Kingdom	:	Plant kingdom
Phylum	:	Phanerogams
Class	:	Monocotyledon
Family	:	Poaceae
Genus	:	Oryza
Species	:	*sativa*

BENTHAM AND HOOKER CLASSIFICATION

- It is natural system of classification.
- Bentham and Hooker were British taxonomists closely associated with Royal Botanical Garden, Kew.
- They published a book "Genera Plantarum" in three volumes in which they identified, named, described and classified 97,205 species which were grouped into 202 orders (now called families).
- While classifying plants they gave importance to free and fused petals and relative position of ovary.

Bentham and Hooker's Classification	Engler and Prantle's Classification
• It is a natural system of classification • Theory of reduction is in operation • Dicots were placed ahead of gymnosperms and monocots • This classification is followed in common wealth countries	• It is a phylogenetic system of classification • Theory of evolution is in operation • Monocots were placed ahead of dicots and gymnosperms • This classification is followed in American and European countries

Out-line of Bentham and Hooker Classification

Kingdom

Out-line of Bentham and Hooker classification

Kingdom:		Plant kingdom	
Division:	Cryptogams		Phanerogams
	(Non-flowering plants)		(Flowering plants)
Class:	Dicotyledonae	Gymnospermae	Monocotyledonae
	(2 cotyledons in seed)	(Seeds not enclosed in fruits and thus naked, Ovary - absent)	(One cotyledon in seed) No order 7 series & 34 families
Subclass:	Polypetalae (Petals separated)	Gamopetalae (Petals fused)	Monochlamydae (Petals and sepals are not distinguished and called perianth)
Series:	**1. Thalamiflorae** Flowers have conical thalamus 6 orders & 34 families	**1. Inferae** (Flowers are epigynous with inferior ovary 3 orders & 9 families)	
	2. Disciflorae Flowers have disc thalamus 4 orders & 23 families	**2. Heteromerae** (Flowers are hy-pogynous, superior ovary with more than 2 carpels 3 orders & 12 families)	
	3. Calyciflorae Flowers have cup shaped thalamus 5 orders & 27 families	**3. Bicarpellatae** (Flowers are hypogy-nous, superior ovary with 2 carpels 4 orders & 24 families)	

Author Citation

A name cannot be complete without an author's name. The author's name is abbreviated, e.g., Linneaus is abbreviated as Linn or L, Benthm as Benth; Hooker as Hook, Roxburgh as Roxb, Lamark as Lamk etc. The indications of name of a taxon are to be accurate and complete. It is necessary to cite the name of the author who first validly published the name. If the author's name is too long it should be abbreviated. e.g., *Gossypium hirsutum* L., *Solarium nigrum Linn* etc.

When a genus or taxon of a lower rank is altered in upper rank but retains its name or epithet, the author who first published this as a legitimate name or epithet must be cited in parentheses; followed by the name of the author was effected the alternation e.g., *Cirtus aurantium var. grandis* L; when raised to rank of species it become *Citrus grandis* (L) Obseck. Here L is the first author and Osbeck altered it.

AGRICULTURAL CLASSIFICATION OF CROPS

1. **Cereals and Millets:** The word cereal is derived from 'Ceres' which means Roman Goddess of Harvest. Larger grain of poaceae is called cereals. Eg. Rice, Wheat, Maize.

   Small grain of poaceae is called Millets. Eg. Sorghum, Pearlmillet. Very small grain of poaceae is called minor millets / small millets. Eg. Ragi, Thenai, samai,
2. **Pulses:** The term pulse is used to denote the seeds of leguminous plants. They are rich in protein and form chief source of protein in vegetarian diet. Eg. Blackgram, Greengram, Redgram, Cowpea.
3. **Oil seeds:** They are the sources of vegetable oils and give taste to the food. Eg. Groundnut, Gingelly.
4. **Fibers:** Fibers are obtained from these crops. Eg. Cotton, Jute, Mesta, Sunnhemp.
5. **Sugars and Starches:** Sugars act as sweetening agent. Eg. Sugar cane, Sugar beet.

   Starches are not sweetening agent but they are polysaccharides of simple sugar. Eg. Tapioca.
6. **Vegetables, Fruits and Nuts:** They supply vitamins and minerals. Eg. Cucurbits, Apple, Cashew.
7. **Beverages:** These are drinks which have stimulating effects upon drinking. Eg. Coffee, Tea.
8. **Narcotics, Fumitories and Masticatories:**

   Narcotics: They produce stimulating effects up on placing on tongue. Eg. Opium.

   Fumitories: They produce stimulating effects up on smoking. Eg. Tobacco.

   Masticatories: They produce stimulating effects up on chewing. Eg. Betelvine.
9. **Spices and Condiments**

   Spices: They are added in small quantity in food for flavour & taste. Eg. Pepper, Cardamon.

   Condiments: They are added in small proportion in food to add flavour & taste. Eg. Onion, Garlic.
10. **Rubber:** They provide raw material for manufacture of rubber. Eg. Rubber.
11. **Forages:** Forages provide feed for domestic animals. Fodders are plant materials obtained after cutting the forage plant.
    - **Forage grass:** Cumbu Napier Hybrid.
    - **Forage legume:** Alfalfa
    - **Tree fodder:** Agathi

12. **Green manures and Green leaf manures**

- **Green manure:** They are grown *in situ* in the field and ploughed *in situ* to decompose as manure. Eg. Sunnhemp, Dhaincha.
- **Green leaf manure:** They are grown outside the field and the leaves & toppings are put in the field to decompose as manure. Eg. Neem, Gliricidia.

Chapter

2

Plant Morphology / Phytography – Vegetative Morphology

Life span, habit, root, stem, leaf – petiole, leaf margin, leaf apex, leaf shape, venation and phyllotaxy, Modification of roots and leaf; Floral morphology: Kinds of bracts, inflorescence, Structure of flower, androecium, gynoecium, placentation, types of fruits

Morphology: Study of external forms of plant bodies.

Based on duration / life span, plant are classified into

1. **Annuals:** Plants complete their life cycle within a year or a season Eg. Rice, Blackgram.
2. **Biennials:** They require two years or seasons to complete their life cycle Eg. Tapioca, Cabbage, Radish.
3. **Perennials:** They live more than two years and flower every year (Polycarpic) Eg. Mango.
4. **Multiennials:** They live for many years like perennials but do not flower every year. They put forth flowers and fruits only once during their life cycle (Monocarpic) Eg. Agave, Palm.

Based on growth habitplants are classified into:

1. **Trees:** Trees are those which bear woody main trunk and numerous branches Eg. Mango, Neem.
2. **Shrubs:** They are woody plants having no main trunk. Its branches arise from the base forming bushy plant Eg. Nerium.
3. **Herbs:** They are small plant with soft stem Eg. Rice, Pulses.

ROOT

- It develops from radical of the seed.
- Grows against sunlight and positive geotropic.
- They do not have nodes and internodes.

Types of root

1. Tap root
2. Fibrous/Adventitious root
3. Tuberous root

Tap root	Fibrous / adventitious root	Tuberous root
• Primary root remains the largest root of the plant and continues its growth in a downward direction with all other roots of the plant branching off from it (Fig. 2.1 a). • Characteristics of dicots	• Primary root is replaced by a cluster of secondary roots that arise not from the radical but from the base of the stem. (Fig. 2.1 b). • Characteristics of monocots	• Root becomes fleshy due to storage of food. (Fig. 2.1 c).

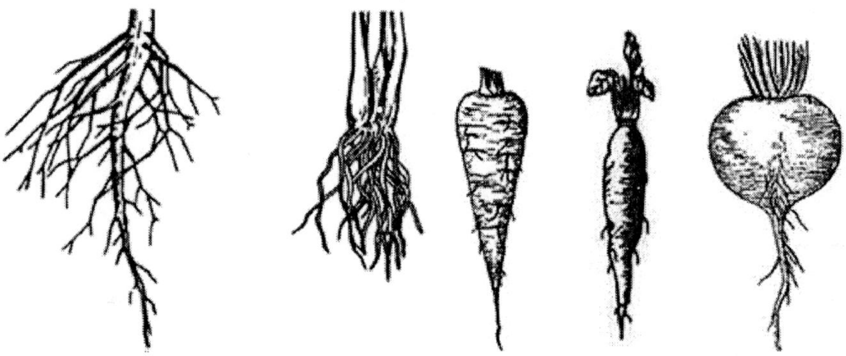

A.Tap root B. Fibrous root C. Tuberous root

Fig. 2.1: Types of Roots: (A) Tap root, (B) Adventitious root C. Tuberous root

Root Modifications:

Tap root Modification

1. **Prop roots:** Hanging aerial roots gives mechanical support to the plant Eg. *Ficus bengalensis* (Fig. 2.2 a).
2. **Epiphytic roots:** Epiphytic plants have special type of roots (Fig. 2.2 b) that grow on other plant in order to have support from them (autotropic). They do not have any relation with soil Eg. *Orchids*.
3. **Pneumatophores:** Special type of roots that arise above water from underground root for respiration (Fig. 2.2 c) in vascular plants growing in sea water or in tidal swamps Eg. *Mangrove*.

Prop root

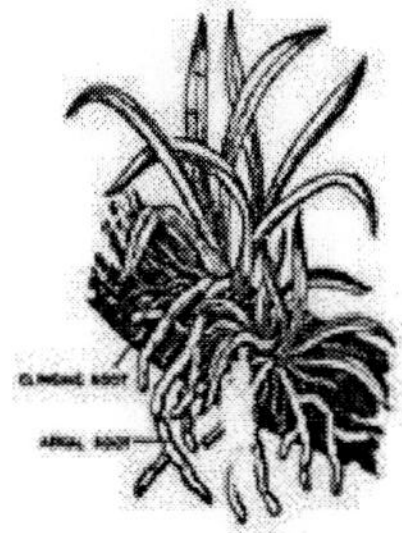

Hanging root

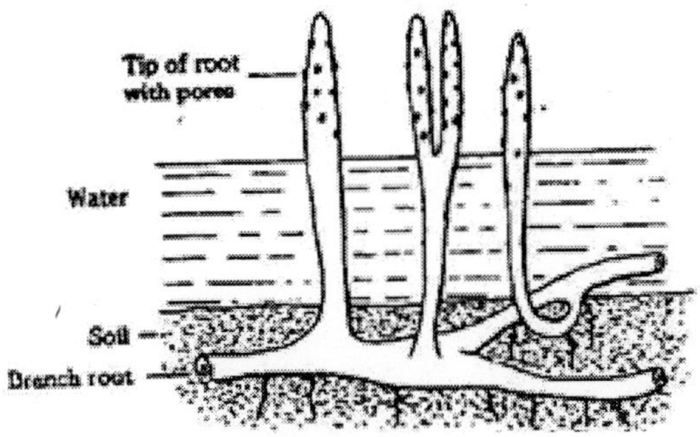

Respiratory root

Fig. 2.2: Tap root Modification

Fibrous root modification:

1. **Stilt root:** Roots arise from nodes of the stem above soil and gives mechanical support to the plant (Fig. 2.3 a) Eg. *Maize, Sugarcane.*
2. **Climbing roots:** Roots that help in climbing of plants (Fig 2.3 b) Eg. *Pepper.*
3. **Respiratory roots:** Spongy roots that arise above ground from underground root for respiration (Fig 2.3 c) Eg. *Mangrone.*

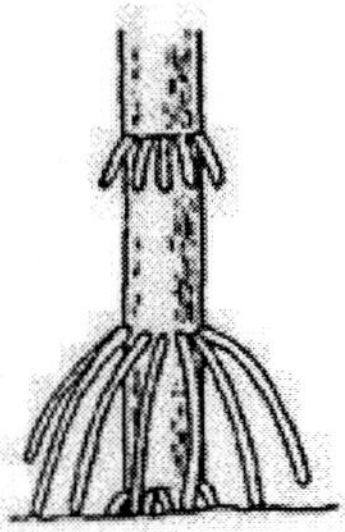

Stilt root

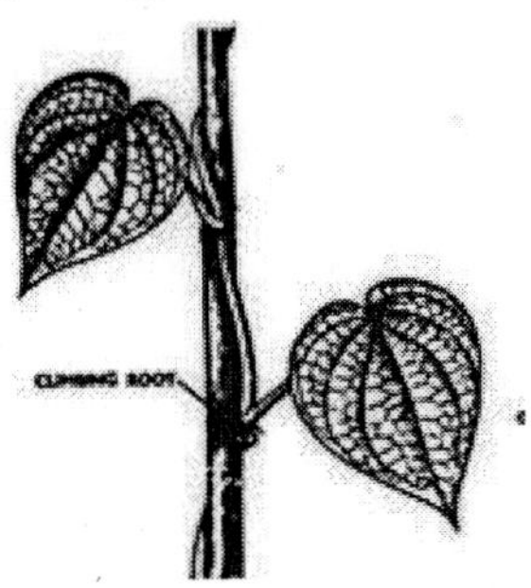

Climbing root

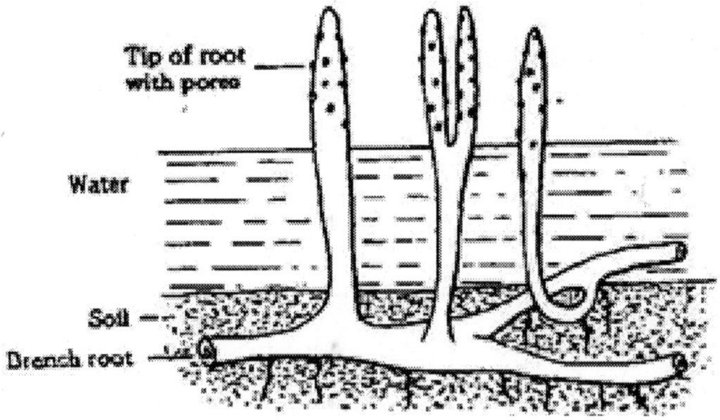

Respiratory root

Fig. 2.3: Fibrous root modification

Tuberous root modification:

Roots become fleshy due to storage of food (Fig. 2.4)

1. **Conicalform:** Cone shaped fleshy root Eg. Carrot.
2. **Fusiform:** Spindle shaped tuberous roots Eg. Radish.
3. **Napiform:** Globose tuberous root Eg. Turnip.
4. **Moniliform:** Beaded swellings of tuberous root Eg. Vitis.
5. **Fasiculated:** Cluster of tuberous root Eg. Asparagus.
6. **Annulated:** Root has a series of ring like swellings Eg. Epicacuhan.

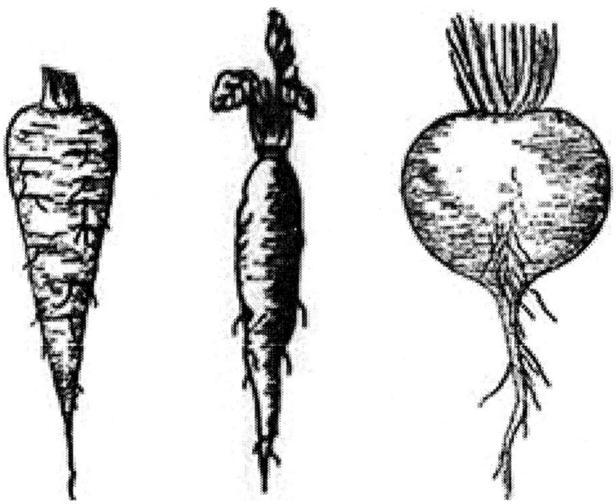

Conical Fusiform Napiform

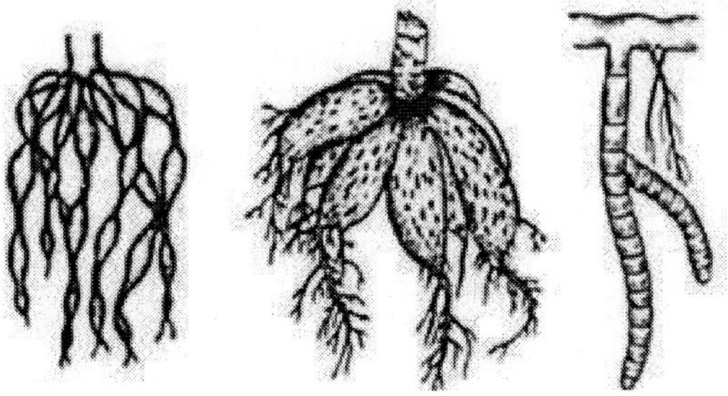

Moniliform **Fasciculated** **Annulated**

Fig. 2.4: Tuberous root modification

Special types of roots

Some roots are modified to give special kind of functions (Fig. 2.5), they are discussed below:

1. **Absorbing roots:** Some plants are epiphytes. These types of roots help in absorption of atmospheric moisture. E.g. Vanda.
2. **Floating roots:** These types of roots are found in aquatic plants and help in floating. E.g. Jussiaea.
3. **Nodulated roots:** These are also the examples of symbiosis in which Bacteria are found in the root nodules of leguminous plants. E.g. Pulses.
4. **Assimilatory roots:** These types are roots have chlorophyll and prepare food for the plants. E.g. Tinospora.
5. **Contractile roots:** In underground stem of some plants like Onion, Garlic etc., some roots are thick and fleshy. Their apical part contracts and fixes the plant in the soil.
6. **Reproductive roots:** Some roots help in reproduction, hence called reproductive roots. E.g. Sweet potato.
7. **Sucking roots:** These types of roots are found in parasitic plant and help in absorption of food from host. E.g. *Cuscuta*.
8. **Mycorrhizal roots:** There is beneficial mutual partnership between a fungus and a root. So it is type of symbiosis. E.g. Pinus and Monotropa.
9. **Butress root:** This type of tree root is formed the trunk which prevent tree fallon and trees found nutrient poor tropical soils.

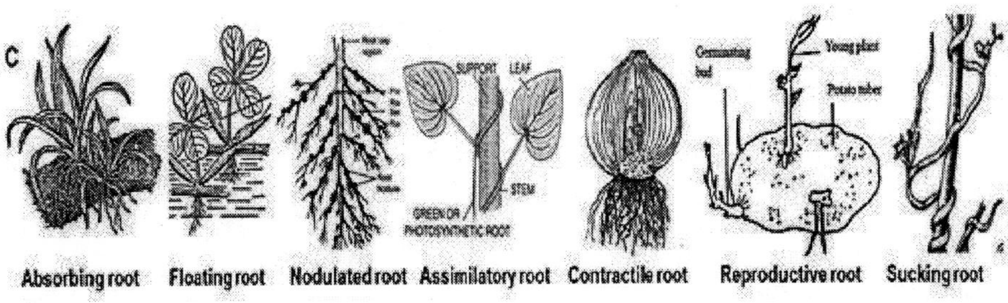

Mycorrhizal roots

Buttress root

Fig. 2.5: Special types of roots

STEM

- It develops from plumule of the seed.
- Growing in the direction of sunlight / in positive phototropic.
- Differentiated into nodes and inter-nodes bearing leaves, flowers and fruits.

Forms of Stem

1. **Erect form of stem:** Stems are strong so that plant can stand vertically erect Eg. Trees, Shrubs, Herbs.
2. **Weak form of stem (Figs.2.6 a,b,c):** Stems are not strong enough to keep themselves vertically erect.
   (a) **Creepers:** Stem or its branches run over the soil surface in all direction and rooting at the nodes giving off daughter plants Eg. *Rangoon creeper*.

Fig. 2a: Creepers

(b) Climbers: Weak slender stem that climb up with the help of some support Eg. *Clitoria*.

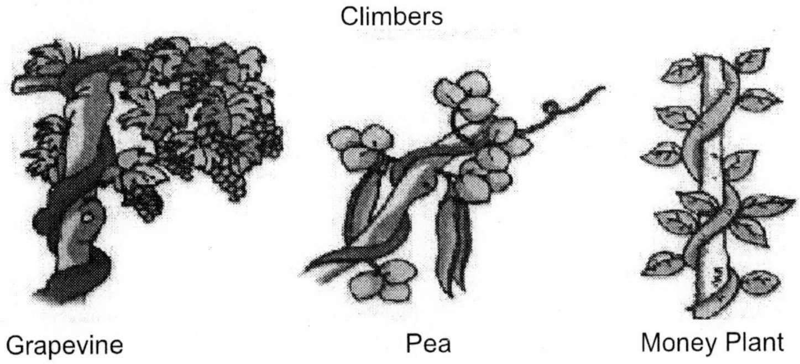

Fig. 2b: Climbers

(c) Trailers: Weak slender stems trail on soil surface. They neither require any support to climb up nor root at the nodes Eg. *Tridax, Nynctanthus*.

Fig. 2c: Trailers

Stem Modifications:

1. Aerial stem modifications(Fig.2.7)

(a) Tendrils: Stem is modified into tendril.

(b) Cladode: Phyllaclade of single inter node Eg. *Asparagas*.

(c) Phylloclade: Stem is modified into flattened and swollen structure which has appearance and function of leaf Eg. *Opuntia*.

(d) Thorns: Stem is modified into spine.

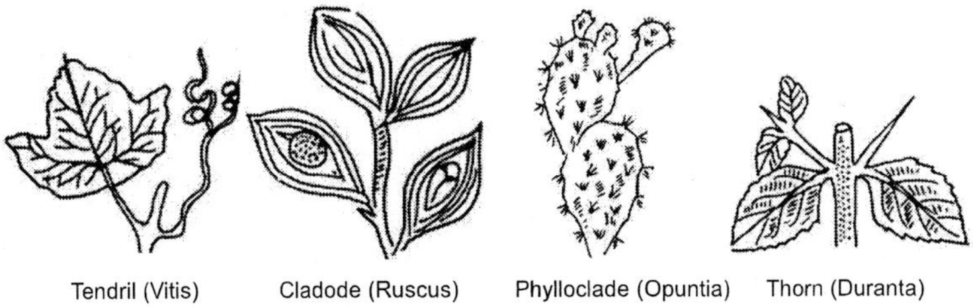

Fig. 2.7: Aerial stem modifications

2. **Subaerial stem modifications (Fig. 2.8):** Meant for vegetative propagation.
   - (a) **Runner:** Plant shows creeping habit of stem run over the soil surface, rooting at nodes produces daughter plants Eg. *Cyanodon dactylon.*
   - (b) **Stolon:** It grows vertically then arches down to contact the soil and produces daughter plants Eg. *Rose.*
   - (c) **Sucker:** It is underground runner. When it comes up above the soil it produces daughter plants. Eg. *Crysanthemum.*
   - (d) **Offset:** It is shorter and thicker runner Eg. *Eichhornia.*

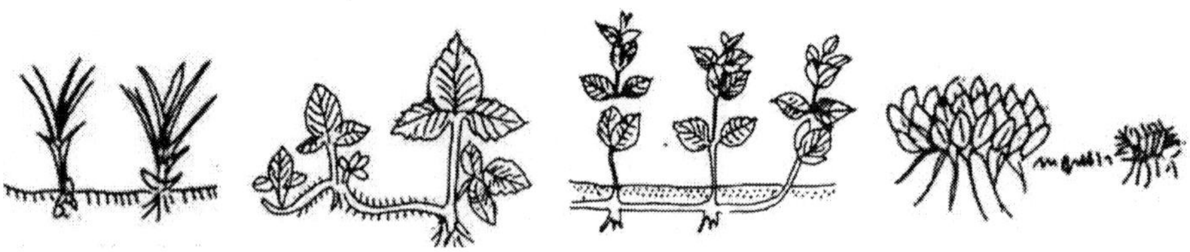

Runner (Cynodon) Stolen (Strawberry) Sucker (Mint) Offset (Pistia)

Fig. 2.8: Subaerial stem modifications

3. **Underground stem modifications(Fig.2.9)**

   Meant for storing the reserved food and vegetative propagation.
   - (a) **Rhizome:** Underground stem distinguished from a root by the presence of nodes, buds and leaves / scales Eg. Turmeric, Ginger.
   - (b) **Tubers:** Swollen underground stem that acts as a storage organ and vegetative propagule Eg. Potato.
   - (c) **Corm:** Swollen underground stem Eg. Colocasia.
   - (d) **Bulb:** Underground storage organ with shortened stem bearing fleshy leaf / scale leaves enclosed the next years' bud. Eg. Onion.

Rhizome (Ginger) Tuber (Potato) Corm (Gladiolus) Bulb (Onion)

Fig. 2.9: Underground stem modifications

LEAF

Parts of Leaf (Phyllopodium) (Fig. 2.10)

- **Leaf lamina / Blade (Epipodium):** Expanded part of a leaf.
- **Petiole (Mesopodium):** Stalk which connects leaf lamina with stem.
- **Leaf base (Hypopodium):** Part of leaf which is attached to the stem.
- **Stipule:** Leaf base bears two small lateral outgrowth called stipules.

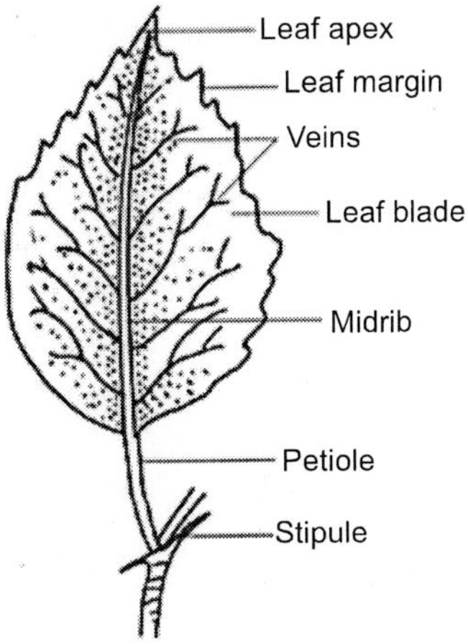

- **Petiolate:** Leaf has petiole.
- **Sessile:** Leaves do not have petiole.
- **Stipulate:** Leaf has stipule.
- **Exstipule:** Leaf does not have stipule.
- **Pulvinus:** Leaf has swollen leaf base.
- **Sheathing leaf base:** Leaf base is broad and flat which clasps the stem as in monocots.

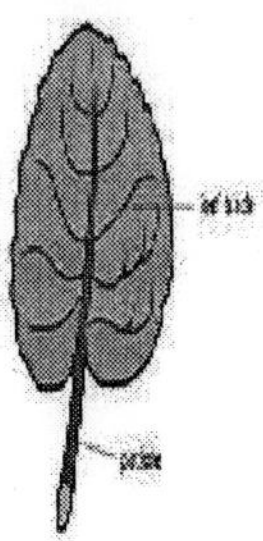

Petiolate

Sessile

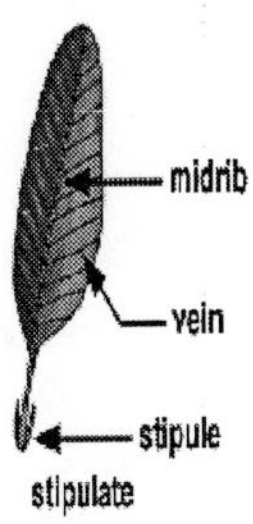

Stipulate

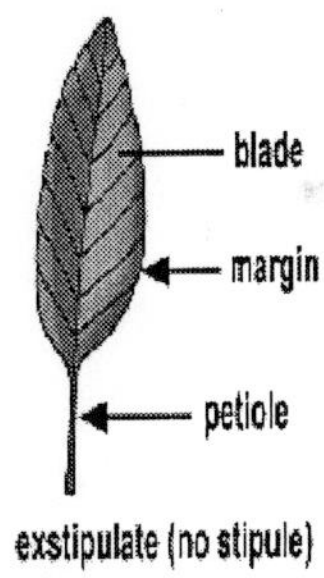

Exstipule

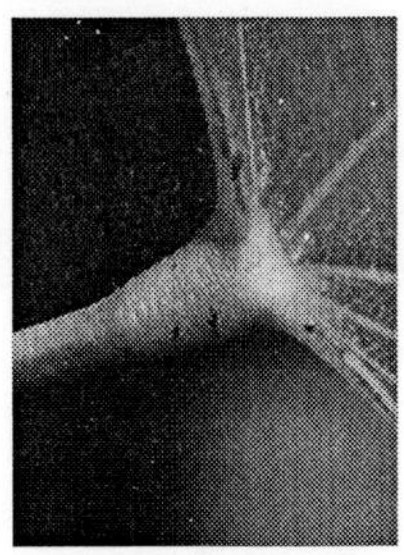

Pulvinus

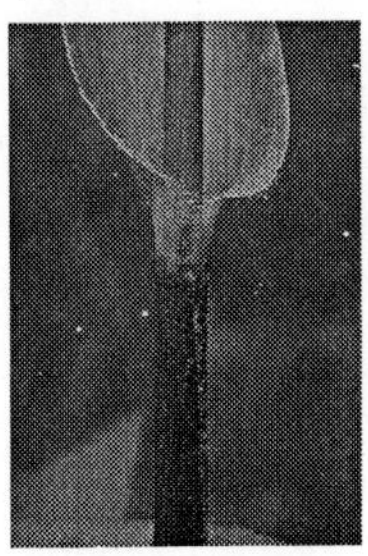

Sheathing leaf base

Fig. 2.10: Parts of Leaf (Phyllopodium)

Types of leaf (Fig. 2.11)

(i) **Simple leaf:** Only one leaf blade attached to the petiole.

(ii) **Compound leaf:** Petiole has number of leaf blades, each having a small stalk of its own. Main petiole is called rachis, each leaf blade is called leaflets / pinna. Compound leaf is of two types:

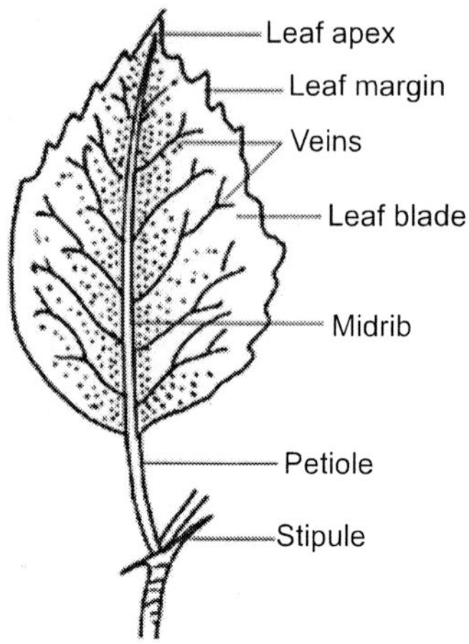

Simple leaf

Compound leaf

Fig. 2.11: Types of leaf

Two types of compound leaf (Fig. 2.12)

1. **Pinnately compound leaves:** Leaflets are borne directly along the length of rachis. Different types are discussed below.

   (a) **Unipinnate:** Leaflets borne directly on the rachis. If number of leaflets is odd and rachis ending with single leaflet is called imparipinnate (Eg. Neem). If number of leaflets is even and rachis ending with pair of leaflet is called paripinnate (Eg. *Tamarind*).

(b) **Bipinnate:** Primary rachis is branched to produce secondary rachis which bears the leaflets Eg. *Acacia*.

(c) **Tripinnate:** Primary rachis is branched to secondary and tertiary rachis which bears the leaflets Eg. *Moringa*.

(d) **Multipinnate/Decompound:** Primary rachis is branched more than thrice. Eg. *Coriander*.

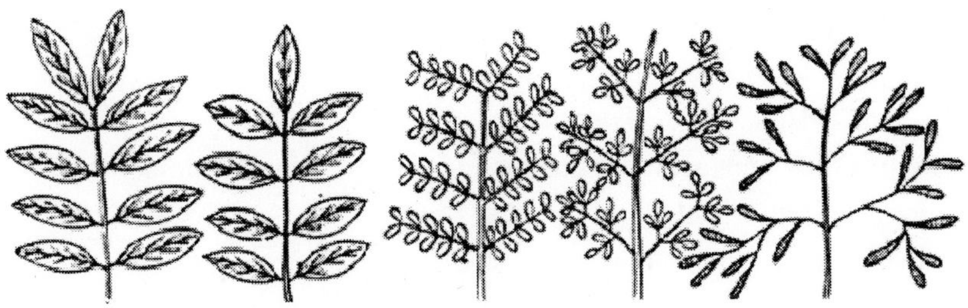

Fig. 2.12: Two types of compound leaf

1. **Palmately compound leaves (Fig.2.13):** Leaflets are attached at a common point at the tip of rachis Different types discussed below:
   (a) **Unifoliate:** Single leaflet attached to the winged petiole Eg. *Lemon*.
   (b) **Bifoliate:** Two leaflets attached to the tip of petiole. Eg. *Zornia*.
   (c) **Trifoliate:** Three leaflets attached to the tip of petiole. Eg. *Pulses*.
   (d) **Quadrifoliate:** Four leaflets attached to the tip of petiole. Eg. *Marsilia*.
   (e) **Multifoliate/Ditate:** More than four leaflets attached to the tip of petiole. Eg. *Bombax*.

Fig. 2.13: Palmately compound leaves

Shape of leaf lamina (Fig. 2.14)

1. **Acicular:** Needle shaped leaf Eg. *Pinus, Onion, Pinus roxburghii*.
2. **Linear:** Long and narrow leaf Eg. *Grasses*.
3. **Lanceolate:** Lance-shaped, broad and tapering either at both the ends or generally towards the apex, e.g., *Nerium indicum*.

4. **Subulate:** It is an awl -shaped leaf, e.g., *Isoetes*.
5. **Oblanceolate:** Broad in the middle and tapering towards both the ends, e.g., *Gnaphalium*.
6. **Elliptical:** Ellipse-shaped with both the ends rounded, e.g., *Psidium, Carissa*, etc.
7. **Ovate:** Egg shaped, i.e. broad at the base and narrowing towards apex, e.g., *Hibiscus rosasinensis*.
8. **Obovate:** It is inversely ovate, i.e., upper end broader narrowing towards the base (Fig. 70), e.g., *Artocarpus*.
9. **Oblong:** Long, wide with parallel-running margins, e.g., *Banana*.
10. **Cordate:** Heart-shaped, e.g., *Piper betel*.
11. **Obcordate:** Inversely cordate, e.g., *Oxalis*.
12. **Round:** When blade is circular, e.g., *Lotus*.
13. **Reniform:** Kidney-shaped (Fig. 70), e.g., *Centella asiatica*.
14. **Spathulate:** Spathula-shaped, i.e., broad and round at the top and narrowing towards the base, e.g., *Drosera*.
15. **Oblique:** When two halves of the lamina are unequal, e.g., *Azadirachta, Melia*, etc.
16. **Sagittate:** When leaf blade is like an arrow (Fig. 70), e.g., *Sagittaria sagittifolia*.
17. **Hastate:** When two basal lobes of a sagittate leaf become pointed and directed outward, e.g., *Ipomoea*.
18. **Cuneate:** Wedge-shaped leaf with lower narrow end, e.g., *Pistia stratiotes*.
19. **Lyrate:** When the leaf blade is like a lyre, i.e., contains a big terminal lobe and many smaller lateral lobes, e.g., *Raphanus sativus*.
20. **Falcate:** When the shape is like a sickle (Fig. 70), e.g., *Eucalyptus globulus*.
21. **Pedate:** Like the claw of a bird, e.g., *Vitis pedata*.
22. **Lorate:** When the shape is like a narrow strip of leather, e.g., *Vallisneria*.
23. **Auriculate:** When the leaf blade is ear-shaped, e.g., *Magnolia fraseri*.

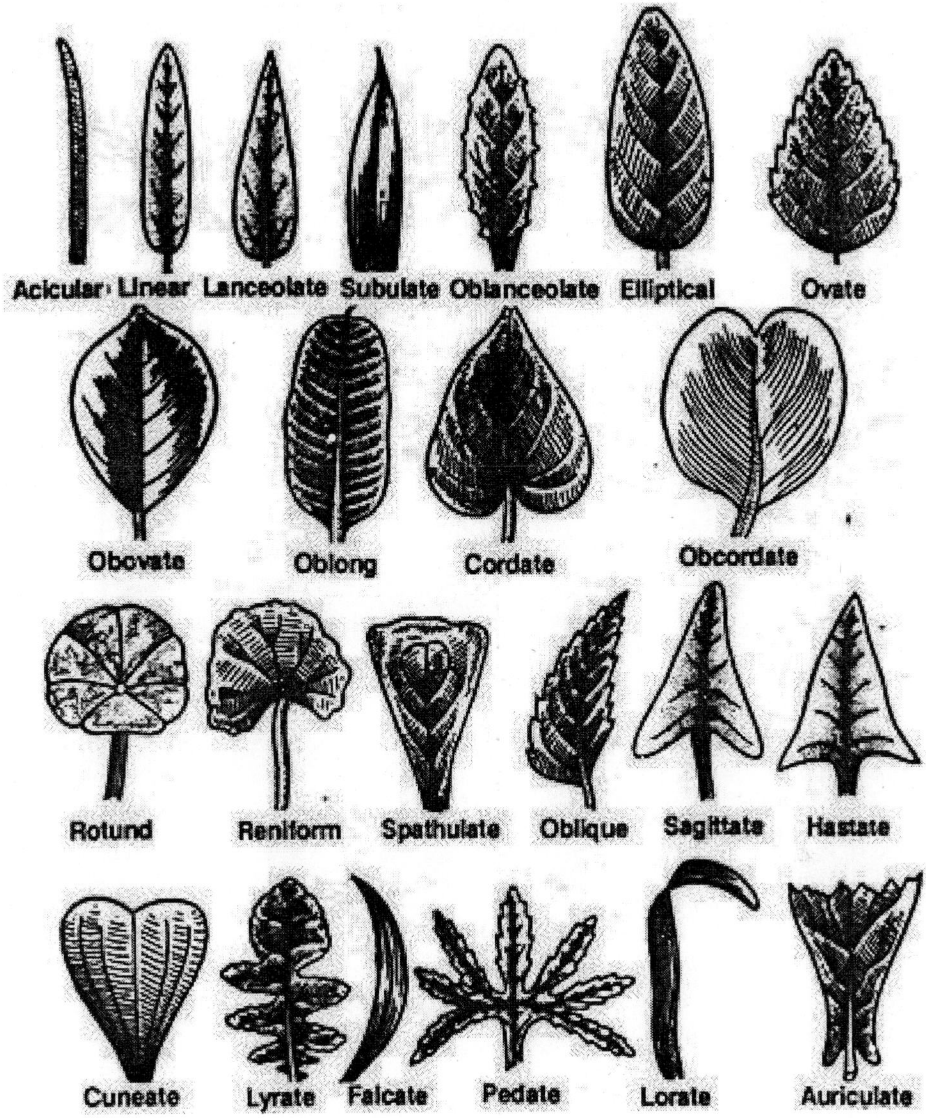

Fig. 2.14: Shape of leaf lamina

Leaf Margin(Fig.2.15)

1. **Entire:** Even and smooth, e.g., *Ficus benghalensis.*
2. **Repand:** Wavy in appearance, e.g., *Polyalthia longifolia.*
3. **Serrate:** Sharp, small teeth – like outgrowths pointing upward, e.g., *Rosa.*
4. **Biserrate:** When each tooth of the serrated margin is again serrated.
5. **Serrulate:** When the teeth in serrate margins are very minute, e.g., *Croton.*

6. **Dentate:** When teeth project outwards at right angle to the edge, e.g., *Nymphaea nouchali.*
7. **Runcinate:** When teeth of the serrated margins are pointing backwards.
8. **Crenate:** When teeth are rounded, e.g., *Bryophyllum.*
9. **Spinous:** Provided with projecting spines, e.g., *Argemone.*
10. **Incised:** Provided with sharp and irregular incisions.
11. **Lobed:** Provided with many lobes, e.g., *Ranunculus.*
12. **Ciliate:** Provided with fine projecting hair, e.g., *Corchorus olitorius.*

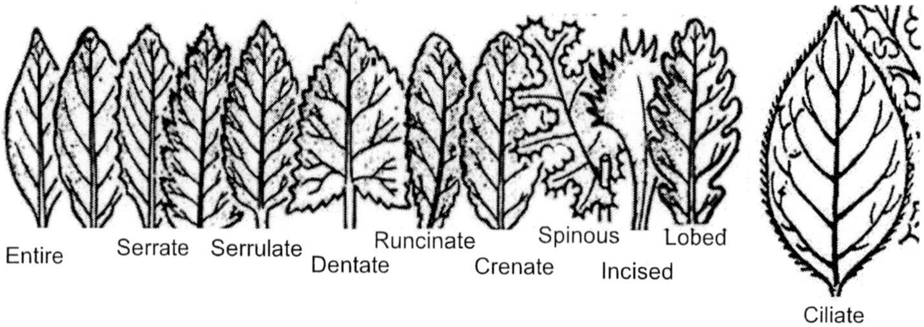

Fig. 2.15: Leaf Margin

Leaf Apex(Fig.2.16)

1. **Mucronate:** Rounded apex with a sharp point Eg. *Cassia.*
2. **Cuspidate:** Spinous apex Eg. *Agave.*
3. **Retuse:** Apex slightly notched Eg. *Oxalis.*
4. **Emarginate:** Having a deep notch at apex Eg. *Bauhinia purpurea.*
5. **Truncate:** Leaf apex appearing cut across abruptly Eg. *Caryota urens.*
6. **Acuminate:** Apex with long tapering tail Eg. *Ficus religiosa.*
7. **Acute:** Pointed apex Eg. *Mango.*
8. **Obtuse:** Rounded apex Eg. *Guava, Banyan.*
9. **Round**
10. **Apiculate:** Forming a small tip at leaf apex Eg. *Dalbergia.*
11. **Cirrhose:** Leaf with prolongation of midrib to form tendril Eg. *Gloriosa.*
12. **Aristate:** Leaf apex found with awns.

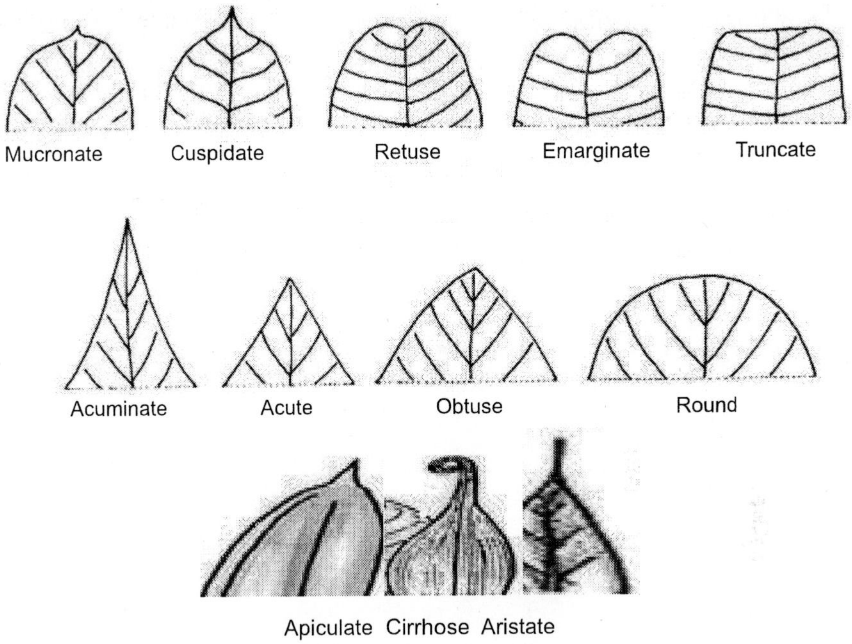

Fig. 2.16: Leaf Apex

Leaf Venation (Fig. 2.17)

Arrangement of veins in leaf blade is called venation.

Two types:

1. Reticulate venation
2. Parallel venation

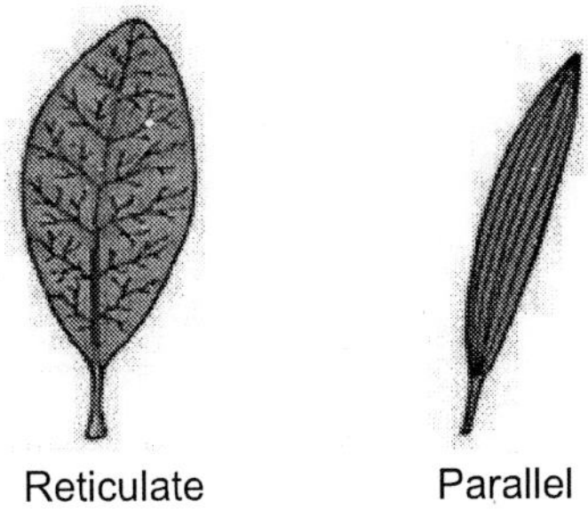

Fig. 2.17: Leaf Venation

1. **Reticulate venation (Fig.2.18):** There is prominent vein called midrib from which many veins arise and from a net like structure. It is common in all dicots.
   (a) **Pinnately reticulate:** One midrib in the center of leaf and many lateral branches to form a net Eg. *Mango*.

(b) **Palmately reticulate:** Several large veins of equal size arise from tip of petiole and many lateral branches to form a net Eg. *Cucurbits*.

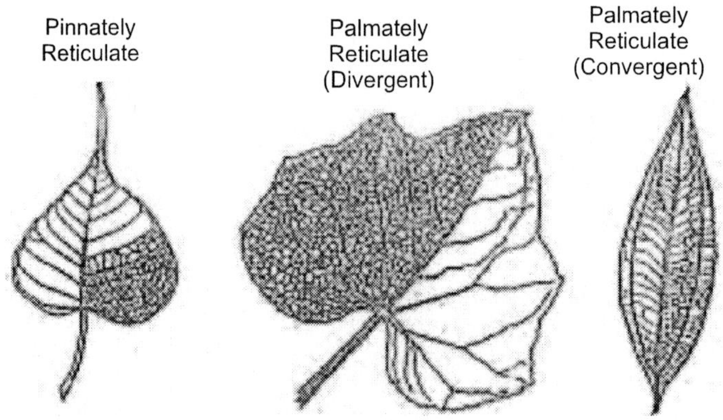

Fig. 2.18: Reticulate venation

2. **Parallel venation(Fig.2.19):** All the veins run parallel to each other. It is common in all monocots.

   (a) **Pinnately parallel:** Only one prominent midrib in the center from which many veins arise perpendicularly and run parallel to each other either convergent at tip or spread out Eg. *Banana*.

   (b) **Palmately parallel:** Several veins arise from tip of petiole, run parallel to each other either convergent at tip as in grasses (Convergent) or spread out as in palmyra (Divergent).

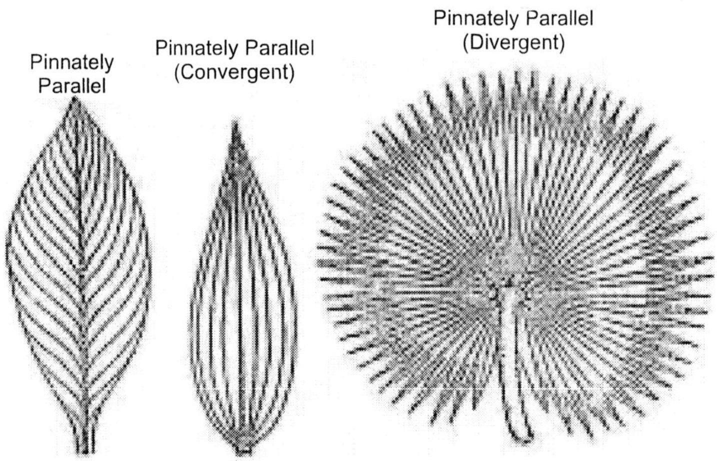

Fig. 2.19: Parallel venation

Phyllotaxy: Arrangement of leaves on stem is called phyllotaxy.

1. **Alternate (Fig.2.20):** Leaves are arranged alternatively on nodes and only one leaf at each node.

   (a) **Distichous:** Alternate leaves are arranged at right and left of the stem Eg. *Grasses*.

(b) **Spiral:** Alternate leaves are arranged spirally on the stem.

Fig. 2.20: Alternate

2. **Opposite (Fig. 2.21):** Leaves are arranged in pairs on nodes lying opposite to each other.

   (a) **Oppositesuperposed:** Pairsofleavesarearrangedineachnodeinsamedirection Eg. *Guava.*

Fig. 2.21: Opposite

   (b) **Opposite decussate(Fig. 2.22):** Pairs of leaves are arranged in each node at right angles to next upper or lower leaf pairs Eg. *Calatropis.*

Fig. 2.22: Opposite decussate

3. **Ternate (Fig. 2.23):** Three leaves are present at each node Eg. *Nerium.*

Fig. 2.23: Ternate

4. **Whorl (Fig. 2.24):** More than three leaves at each node. Eg. *Alamanda.*

Fig. 2.24: Whorl

5. **Radical (Fig. 2.25):** Leaves arise in a cluster from the shortened stem near the ground level Eg. *Agave.*

Fig. 2.25: Radical

Leaf Modifications (Fig. 2.26)

Primary function of leaf is photosynthesis and transpiration (Fig. 2.26). But in some plant modified leaves perform some additional function. Leaves undergo different kinds of modifications in order to perform different functions. These modifications are necessary to adapt to the environmental conditions or to support the growth and survival of the plant itself.

1. **Leaf tendrils:** Leaves are modified into tendrils which help in climbing Eg. *Lathyrus*.
2. **Leaf hooks:** Leaves are modified into hooks Eg. *Bignonia*.
3. **Leaf spines:** Leaves are modified into spines Eg. *Zizyphus*.
4. **Phyllode:** Petiole becomes flattened or winged Eg. *Acacia*.
5. **Pitcher:** Leaves are modified into pitcher which has slender stalk holding the pitcher vertical with a lid. The pitchers capture and digest the insects to get nitrogenous food from them. Eg. *Nepenthes*.
6. **Bladder:** Leaves are modified into bladders Eg. *Utricularia*.
7. **Undulate:** Wavy leaf margin Eg. *Polyalthia*.
8. **Crispate:** Extremely wavy leaf *margin*.
9. **Convolute:** Rolled together.
10. **Lacerate:** Margin cut deeply into irregular lobes.
11. **Lacinate:** Irregularly incised / fringed.
12. **Laciniolate:** Minutely incised / fringed.
13. **Ciliate:** Bearing fine hairs on the leaf margin.
14. **Spinous:** Bearing many spines on leaf margin.
15. **Pectinate:** Comb like leaf margin.
16. **Lobed:** Leaf margin divided into many lobes.
17. **Glandular:** Sticky glandular hairs on leaf margin.
18. **Fid:** Leaf blades cut into half the distance from margin to midrib / top of petiole and forms number of leaf segments.
19. **Partite:** Leaf blades cut into ¾ the distance from margin to midrib / top of petiole and forms number of leaf segments.
20. **Sect:** Leaf blades cut nearly to midrib / top of petiole and forms number of leaf segments.

Prominent leaf modification

1. **Leaf Tendrils:** Tendrils are thread-like green structures and sensitive to touch. Capable of coiling around a support. They can be differentiated from stem tendrils by the absence of scaly leaves over them and unbranched. E.g. Pea.
2. **Leaf Spines:** Leaves are modified into spines. Spines protect the plant from grazing animals. They reduce the area for transpiration hence can prevent water loss under xerophytic conditions. E.g. *Opuntia* (Fig. 2.26).
3. **Food Storage:** Food is stored in the leaves. Leaves may become fleshy or juicy due to storage of water, mucilage and food material. E.g. *Onion* (Fig. 2.26).
4. **Phyllode:** Petiole gets flattened and green in colour to perform the function of photosynthesis. E.g. *Australian acacia* (Fig. 2.26).

5. **Leaf Bladder:** The leaves of some insectivorous plants are modified into bladder like structures. They are very sensitive to touch and have mechanism to catch insects which sits on them. Once they trap the insect, they secrete digestive enzymes inside the bladder to digest the insect. E.g. *Bladder wort* (Fig. 2.26).

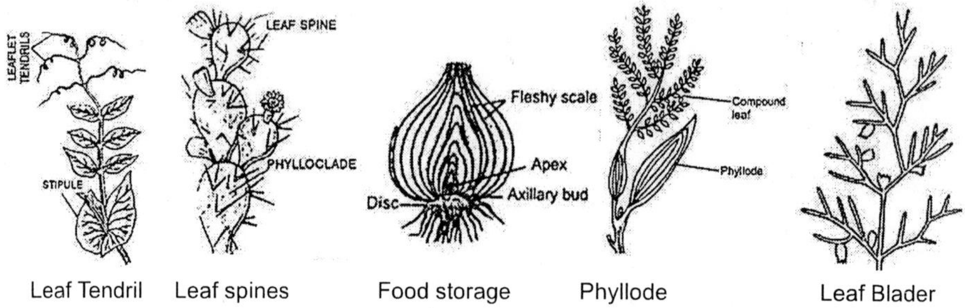

Fig. 2.26: Leaf modifications

Chapter

3

Plant Morphology / Phytography – Floral Morphology

Kinds of bracts, Inflorescence; Structure of flower, Androecium, Gynoecium, Placentation, Types of fruits

INFLORESCENCE

Several flowers arise in a cluster on a common axis called inflorescence. Common axis of inflorescence is called rachis or peduncle. If peduncle arises directly from underground rhizome called scape Eg. Onion, Agave

Types of inflorescence

Based on arrangement of flowers (Fig.3.1) it is classified into:

1. Racemose
2. Cymose
3. Mixed inflorescence
4. Special types of inflorescence

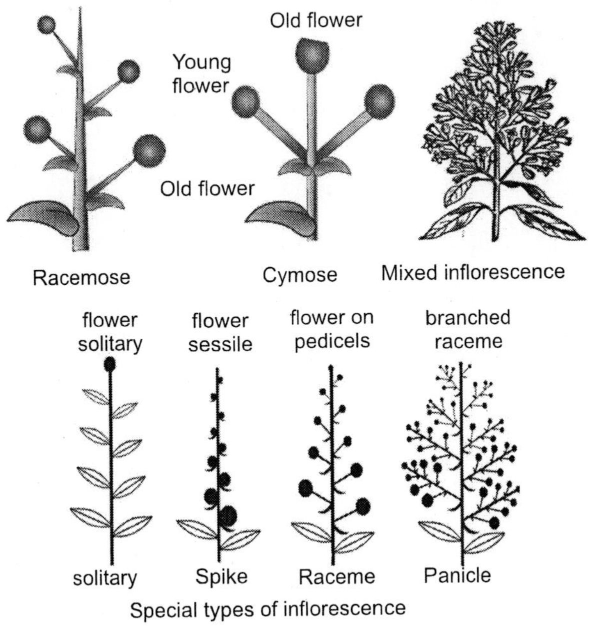

Fig. 3.1: Types of inflorescence

Racemose

This type of inflorescence shows indeterminate growth of inflorescence axis which does not end in a flower. Flowers are arranged in acropetal succession ie., older flowers found on the base and younger flowers found towards tip of inflorescence axis. There are different types:

1. **Main axis elongated:** Inflorescence axis is elongated which bears pedicillate / sessile flowers.
   - **(a) Spike:** This type has unlimited growth of inflorescence axis and sessile flowers are arranged in acropetal succession. Eg. *Pepper*.
   - **(b) Simple raceme:** This type has unlimited growth of inflorescence axis and pedicillate flowers are arranged in acropetal succession. Eg. *Crotalaria*.
   - **(c) Compound raceme(Fig.3.2a):** Inflorescence axis is branched and each branch shows as simple raceme. Eg. *Mango*.

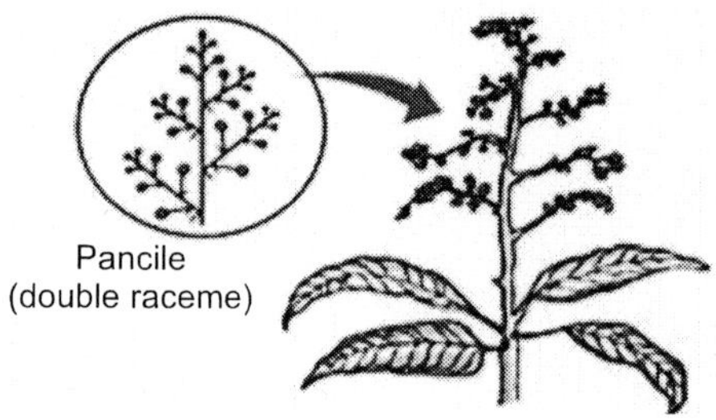

Fig.3.2 a Compound raceme inflorescence

   - **(d) Compound spike:** Inflorescence axis is branched and each branch shows as spike (Branched spike).
   - **(e) Spadix:** Inflorescence axis is swollen and numerous sessile flowers are arranged in acropetal succession. Entire inflorescence is protected by a large bract called spathe. Base of axis bears female flowers while male and sterile flowers towards top. Eg. *Colocasia*.
   - **(f) Compound spadix:** It has branched swollen inflorescence axis. Eg. *Coconut*.

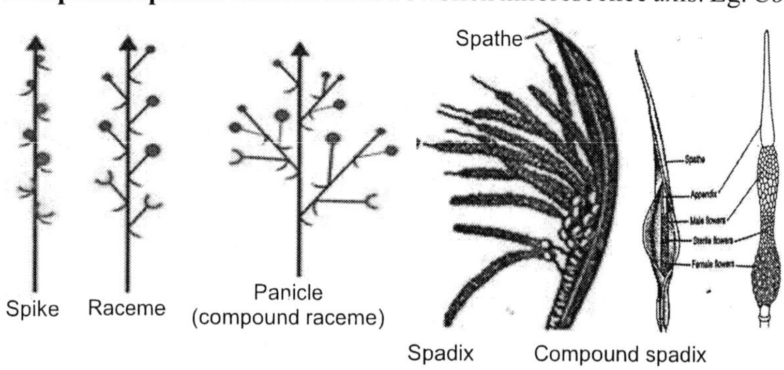

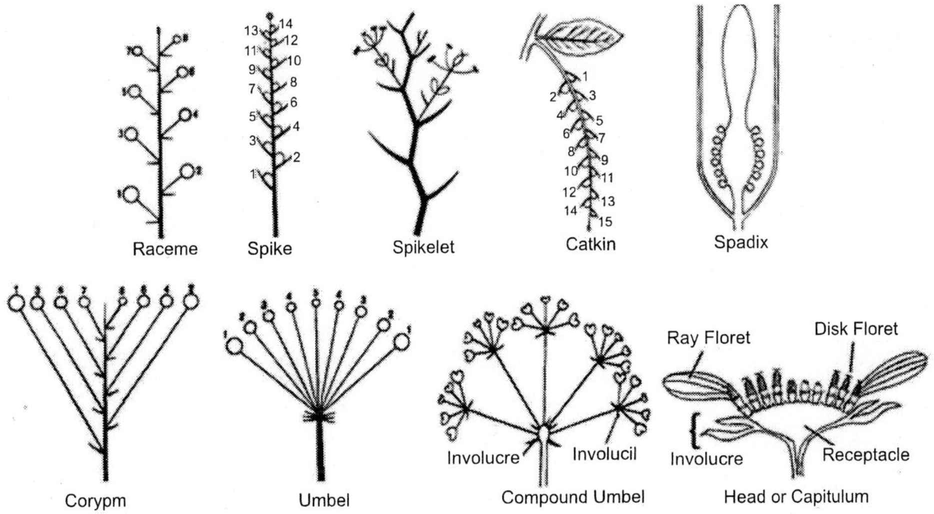

Fig.3.2: Main axis elongated

2. **Main axis shortened:** Inflorescence main axis is reduced

   (a) Corymb: It has reduced inflorescence axis and pedicels of flowers are of unequal length (Fig. 3.3). Older flower has long pedicel and younger shows shorter pedicel. So flowers appear at the same level. Eg. *Caesalpinia.*

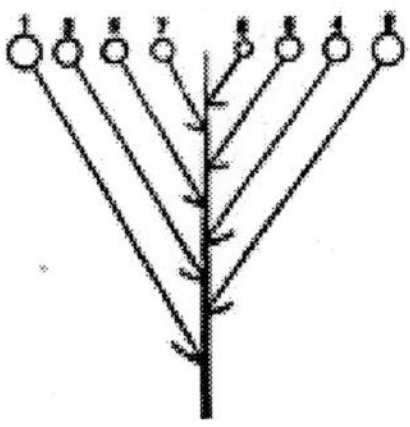

Corymp

Fig.3.3: Corymb

3. **Main axis ending in a flower (Fig. 3.4)**

   (a) **Umbel:** Growth of inflorescence axis is suddenly stopped and a whorl of bracts arise at tip of the inflorescence called involucre of bracts, from which pedicellate flowers are arranged in acropetal succession. Eg. *Onion.*

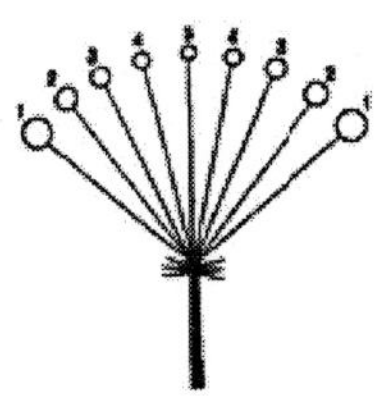

Umbel

Fig. 3.4: Umbel

(b) **Compound Umbel (Fig.3.5):** Inflorescence axis ended with involucre of bracts which gives raise to many branches each produces involucre of bracts from which pedicellate flowers are arranged in acropetal succession. Eg. *Carrot.*

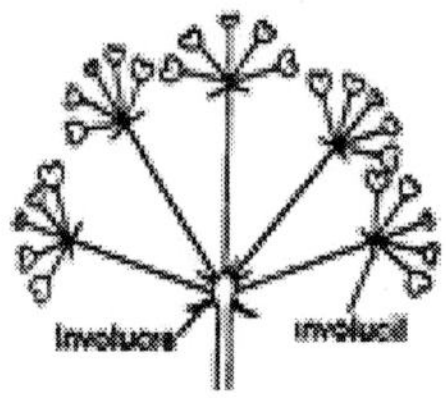

Compound Umbel

Fig.3.5: Compound Umbel

4. **Main axis flattened:** Main axis is flattened on which flowers are arranged.

Two types are discussed below:

(a) **Head / Capitulam(Fig. 3.6):** Main axis is flattened and functions as thalamus on which sessile florets are arranged in acropetal order. This inflorescence is surrounded by bracts called involucre which protects young florets and fruits. There are two types of sessile florets viz., Tubular / Disc florets

Ligulate / Ray florets

Based on type of florets, Head is of two types:

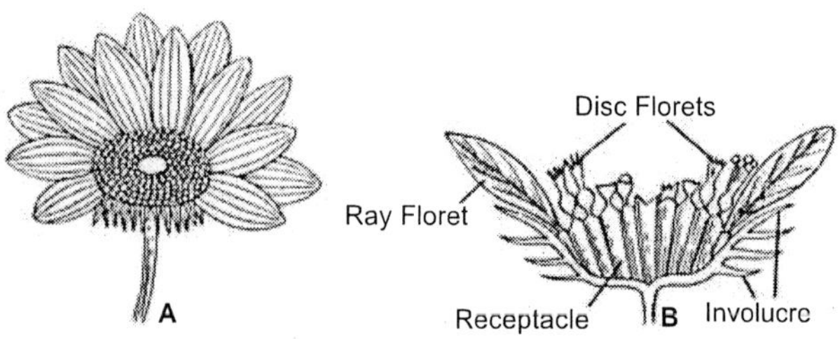

A. Head or Capitulum **B. LS OF head**

Fig.3.6: Head or Capitulam

- **Homogamous Head(Fig.3.7):** Head has a single kind of florets either ray or disc florets.

  Eg. Vernonia has only disc florets;

  Launaea has only ray florets.

Fig. 3.7: Homogamous Head

- **Heterogamous Head (Fig. 3.8):** Head has both kinds of florets. Disc florets are present in the centre of thalamus while ray florets radiate outside from the margin of thalamus Eg. *Helianthus*, *Tridax*.

Fig. 3.8: Heterogamous Head

(b) **Compound Head (Fig. 3.9):** Inflorescence axis is branched and each bears a head. Eg. *Lagasca mollis*.

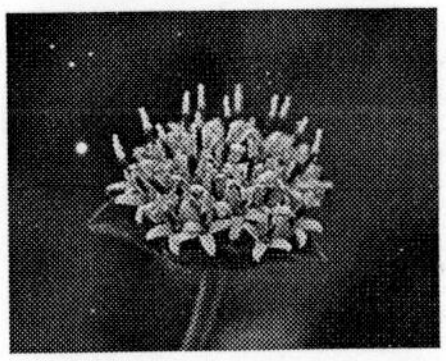

Fig. 3.9: Compound Head

Cymose Inflorescence:

- Inflorescence axis / peduncle has limited growth.
- Tip of inflorescence stops growing by producing a flower and further growth is due to formation of lateral branches that also ends in a flower.
- Flowers are formed in basipetal order ie. From apex to base. Flower opening is in centrifugal ie. From centre to periphery.

Different Types

1. **Simple / Solitary Cyme(Fig.3.10)**

   Main stem or leaf axil has s single flower attached on the pedicel called terminal solitary and axillary solitary cyme respectively.

   Eg. *Papaver has terminal solitary cyme. Hibiscus has axillary solitary cyme*

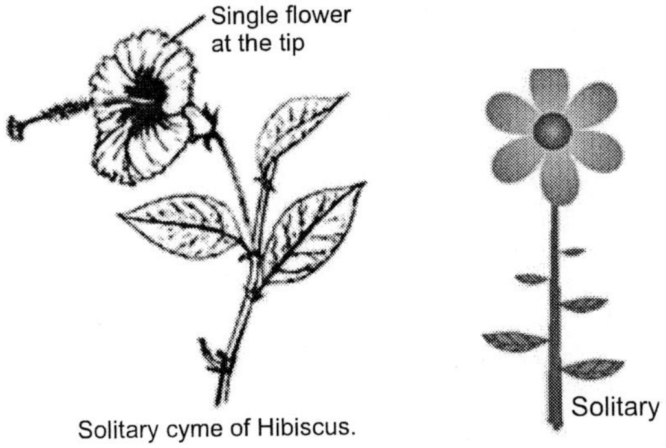

Fig. 3.10: Simple / Solitary Cyme

2. **Simple Dichasium:** Inflorescence axis ends in a flower, two lateral bracts at the base of flower give rise to branches ending in a flower. Thus, there are three flowers in the inflorescence and the central flower is the oldest. Eg. *Jasminum*.
3. **Compound Dichasium:** A branched simple dichasium forms a compound dichasium. Inflorescence axis ends in a flower; from lateral bracts pair of branches arise ending in a flower. From each branch pair of bracts give rise to lateral branches. Eg. *Clerodendron*.
4. **Monochasial Cyme:** Infloerescence axis ends in a flower. From two lateral branches only one branch arise further.
   (a) **Helicoid Cyme:** Inflorescence axis ends in a flower; lateral branches arising from bracts on one side only and thus gives helical appearance Eg. *Hamelia patens*.
   (b) **Scorpioid Cyme:** Inflorescence axis ends in a flower; lateral branches arising from bracts alternately to the left and right in a zig-zag manner. Eg. *Heliotropium*.
5. **Polychasial Cyme (Fig. 3.11):** Main axis ends in a flower; lateral branches arise from the bracts repeatedly. Eg. *Nerium*.

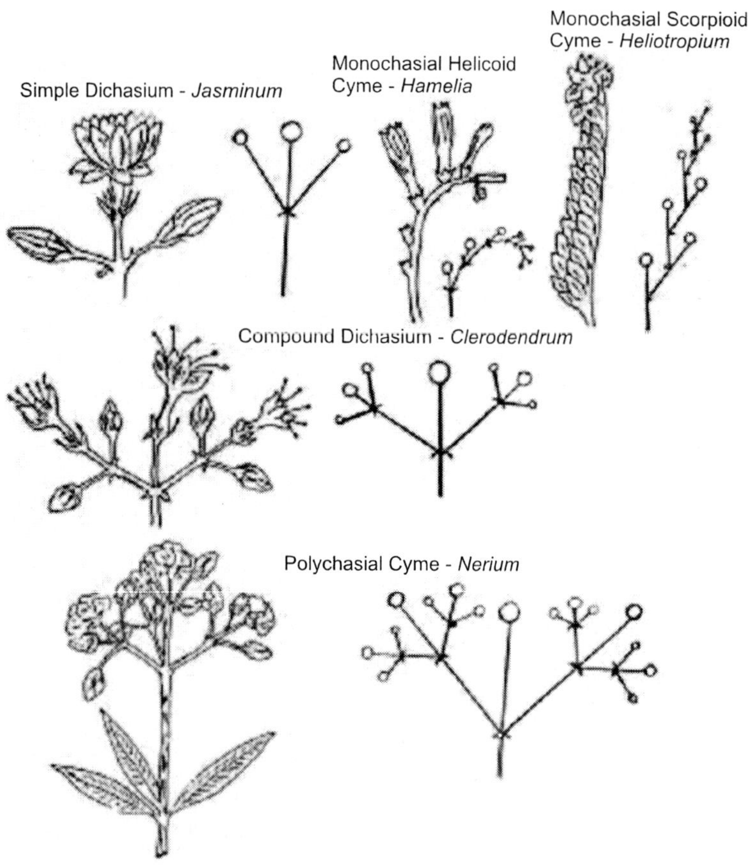

Fig.3.11: Polychasial Cyme

Mixed inflorescence

Inflorescence axis starts as racemose and their branches show cymose fashion.

1. **Thyrsus:** Simple dichasial cyme arranged in a racemose manner in the florescence axis. Eg. *Ocimum*.
2. **Verticillaster:** Pair of dichasial cymes and branched as monochasial scorpioid cyme. Eg. *Leucas*.
3. **Mixed spadix(Fig.3.12):** Serial cymose clusters are arranged on the swollen inflorescence axis from base to apex and each cymose cluster is surrounded by a large bract called spathe Eg. *Musa*.

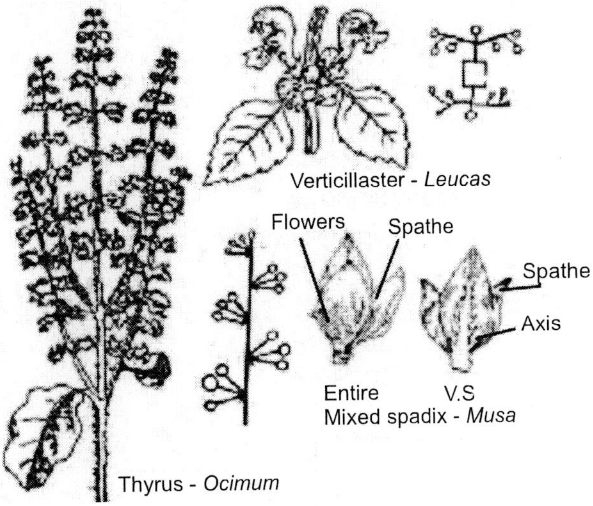

Fig. 3.12: Mixed spadix

Special type of inflorescence

This type of inflorescence can not be included either in racemose or cymose type.

1. **Cyathium:** Inflorescence axis is reduced looks like a single flower; bracts are united to form a cup like structure which encloses receptacle / thalamus. There is a single reduced female flower borne on a large stalk in the centre of receptacle and many reduced male flowers arise in the axil of a bract. Eg. *Euphorbia cyathophora*.

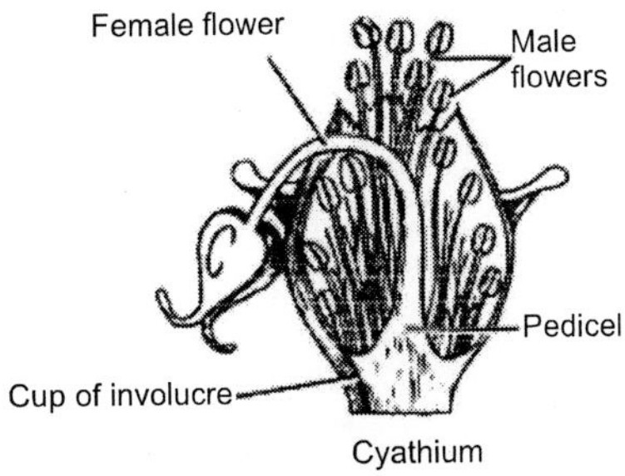

Fig.3.13: Cyathium

2. **Hypanthodium:** The receptacle is concave and cup shaped. The upper end has opening called ostiole (Fig.3.14) which is protected by scales. Inner wall of receptacle has three types of flowers. Male flowers are present in the upper part, female flowers in base and neutral flowers in the middle. Eg. *Ficus*.

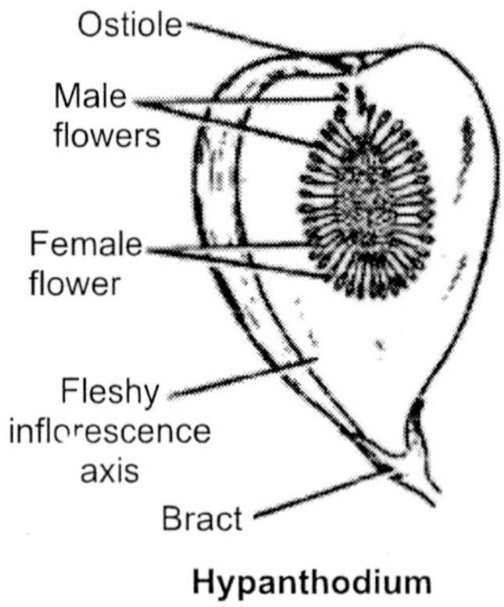

Fig. 3.14: Hypanthodium

3. **Coenanthium:** Receptacle is fleshy and circular disc like structure (Fig. 3.15). Female flowers are present in the centre and male flowers are in periphery of receptacle. Eg. *Dorstenia*.

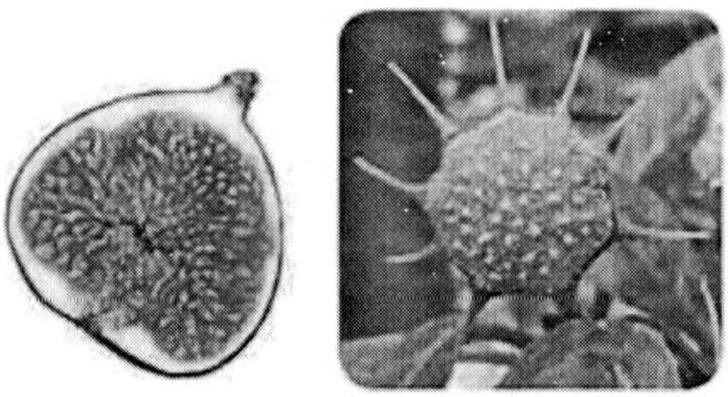

Hypanthoium **Coenanthium**

Fig. 3.15. Coenanthium

Structure of flower

Flower is a modified condensed shoot to carry out sexual reproduction in higher plants (Fig. 3.16).

1. **Terminal flower:** Flower arises in apices of stem.
2. **Axillary flower:** Flower arises in axil of a leaf.
3. **Pedicel:** Flower has a short stalk.
   - **Pedicellate:** Flowers have stalk.
   - **Sessile:** Flowers do not have stalk.

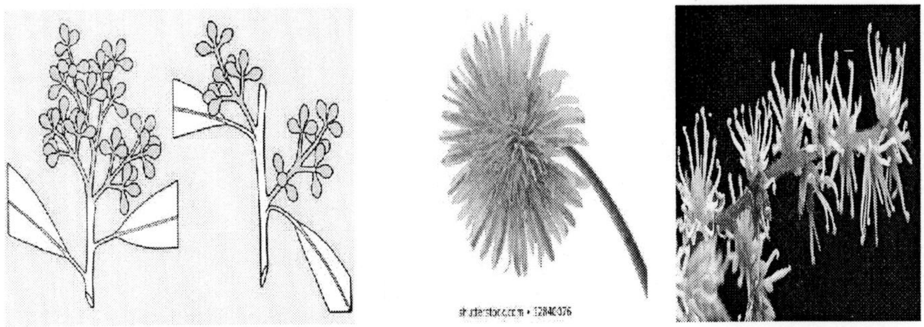

Terminal flower Axillary flower Pedicel flower Sessile flower

Fig. 3.16: Structure of flower

Parts of flower

1. Bracts and Bracteoles
2. Thalamus
3. Whorls of flower (Fig. 3.17)
   (a) Calyx
   (b) Corolla
   (c) Androecium
   (d) Gynoecium

- Calyx and corolla are non-essential / accessory organs, since they do not have direct role in reproduction.
- Androecium and gynoecium are essential organs, since they have direct role in reproduction.

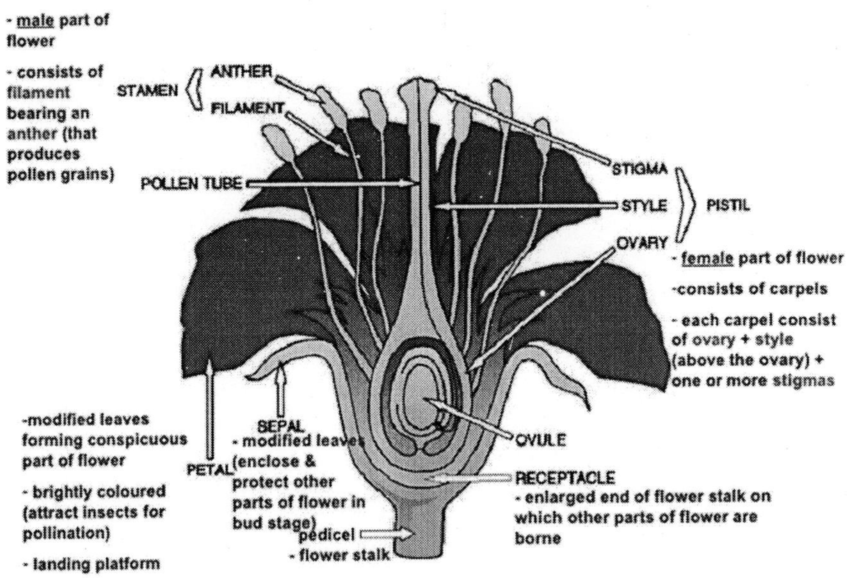

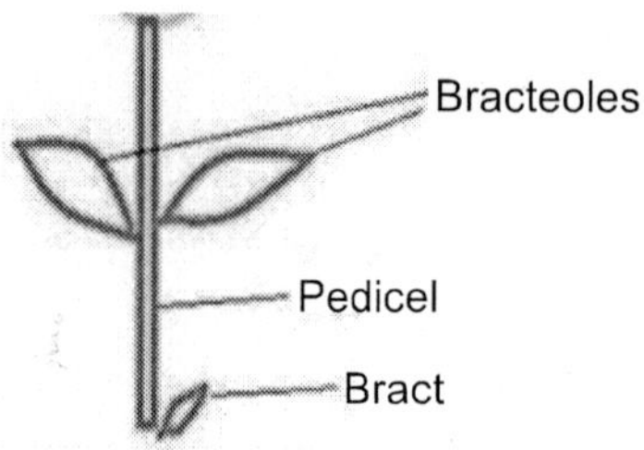

Fig. 3.17: Parts of flower

- **Bracts:** They are special leaves at whose axil flowers develop.
- **Bracteate:** Flowers have bract.
- **Ebracteate:** Flowers do not have bract.
- **Bracteoles:** A small and thin bract like structure present between flower and bract on the pedicel called bracteoles.
- **Bracteolate:** Flowers have bracteoles.
- **Ebracteolate:** Flowers do not have bracteoles.
- **Receptacle/Thalamus:** It is a short floral axis with compressed node and internodes on which floral whorls are inserted.
- **Anthophore:** Elongated internode between calyx and corolla. Eg. *Caryophyllaceae.*
- **Androphore:** Elongated internode between corolla and androecium Eg. *Passiflora.*
- **Gynophore:** Elongated internode between androecium and gynoecium Eg. *Capparis.*
- **Gynandrophore:** Both androphore and gynophores present. Eg. *Gynandropsis.*
- **Carpophore:** Thalamus prolongated beyond the ovary.
- **Hypogyny:** Thalamus is convex. Ovary present in top most position and floral members like sepals, petals and stamens are placed below ovary called superior ovary Eg. *Rice.*
- **Epigyny:** Thalamus is cup shaped. Floral members are inserted above ovary called inferior ovary. Eg. *Helianthus.*
- **Perigyny:** Thalamus is flat / slightly cup shaped. Ovary is situated at centre of thalamus and floral members are inserted on its margin Eg. *Fabaceae.*
- **Perianth:** Calyx and corolla are not differentiated.

Calyx

It is the outermost whorl of a flower which has sepals. They are usually green in colour. Sometimes, calyx becomes coloured called petaloid.

- **Polysepalous:** Sepals are free.
- **Gamosepalous:** Sepals are united.

After opening of flower calyx usually fall off. However, sometimes it may persist.

- **Fugacious / Caducous:** Calyx falls off before flower opening.
- **Deciduous:** Calyx falls off after opening of flower.
- **Persistent:** Calyx persists even after fruit formation.
- **Accrescent:** Calyx grows along fruit development.

Corolla

Second floral whorl which has petals:

- **Polypetalous:** Petals are free.
- **Gamopetalous:** Petals are united.

Function

- To attract insects for pollination.
- To protect essential organs of flower.

Aestivation(Fig.3.18): Mode of arrangement of either sepals/ petals in floral bud is called aestivation.

1. **Valvate:** Sepals / petals meet by their edges without overlapping Eg. *Sepals of Hibiscus*.
2. **Twisted:** One margin of each sepal / petal overlaps next one regularly either in clockwise or in anti clockwise direction. Eg. *Petals of Hibiscus*.
3. **Quincuncial:** Two petals are ineternal and two petals are external and 5th petal has one margin internal and one margin external. Eg. *Guava*.
4. **Imbricate:** One sepal / petal is internal being overlapped on both the margins and one sepal / petal is external or overlapping. Of remaining sepals / petals one margin is overlapped and other overlapping.
   - (a) **Decendingly imbricate / Vexillary:** Overlapping is in descending order. The posterior petal overlaps one margin of two lateral petal and the other margin of two lateral petal overlaps the two anterior petal which are united Eg. *Corolla of Fabaceae*.
   - (b) **Ascendingly imbricate:** Posterior petal is innermost being overlapped by one margin of two lateral petals. Other margin of two lateral petals is overlapped by two anterior petals. Eg. *Petals of Caesalpiniaceae*.

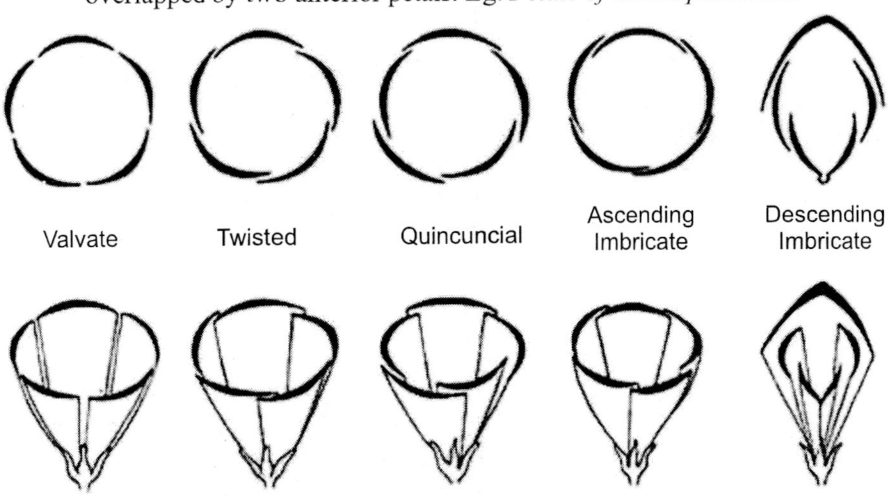

Fig. 3.18: Aestivation of corolla

Androecium

It is third whorl of flower and male part of flower. It is made up of stamens (microsporophylls). Each stamen has stalk (filament) which bears anther (microsporangia). Anther has two lobes connected by connective tissue. Each anther lob has two pollen sacs. Each pollen sac has innumerable pollen grains (microspores). In some plants, a stamen may not produce fertile anther called staminodes.

Based on cohesion of stamens (Fig.3.19): (Union of stamens among themselves).

1. **Monadelphous:** Filaments of stamens are united in one bundle and anthers are free. Eg. *Hibiscus*.
2. **Diadelphous:** Filaments of stamens are united in two bundles and anthers are free. Eg. *Clitoria.*
3. **Polyadelphous:** Filaments of stamens are united in more than two bundles and anthers are free. Eg. *Citrus.*
4. **Syngenesious:** Anthers are united as cylinder and filaments are free. Eg. *Sesame.*
5. **Synandrous:** Anthers and filaments are united through out their length. Eg. *Cucurbitaceae.*
6. **Polyandrous:** Stamens are indefinite and free. Eg. *Ranunculus.*

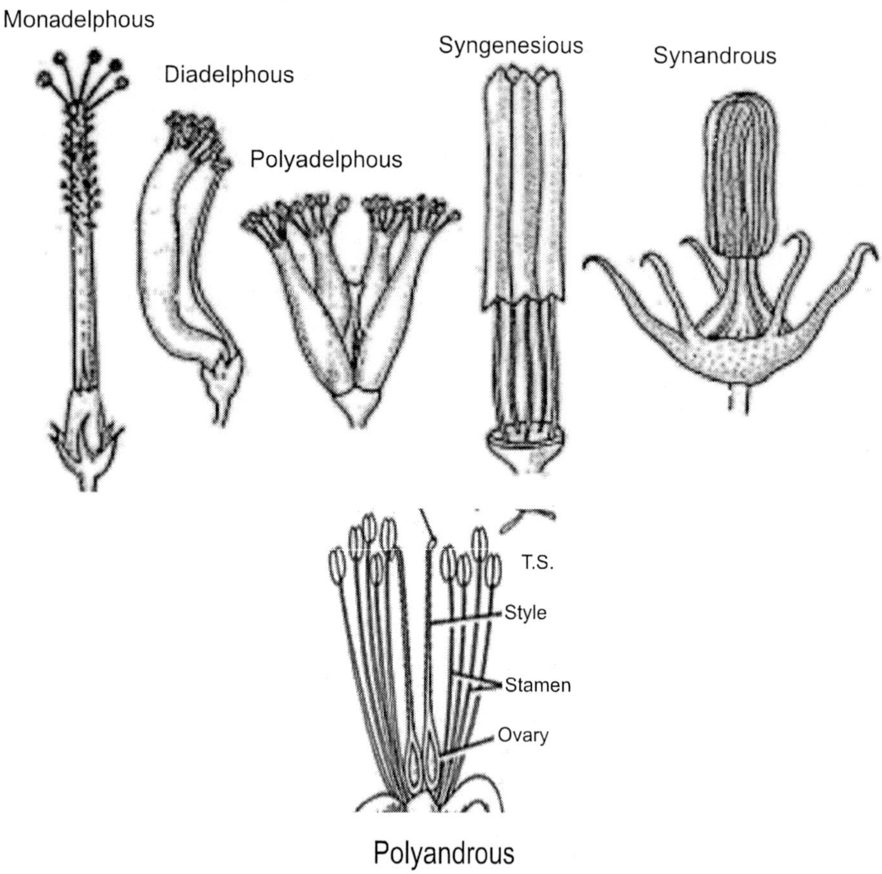

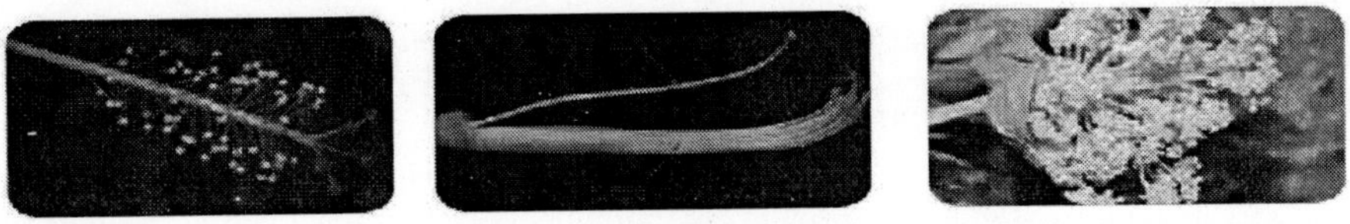

Monadelphous Diadelphous Polydelphous

Fig. 3.19: Based on cohesion of stamens

Based on adhesion of stamens

1. **Epipetalous(Fig.3.20):** Stamens adhere to petals by their filaments. Eg. *Solanum.*

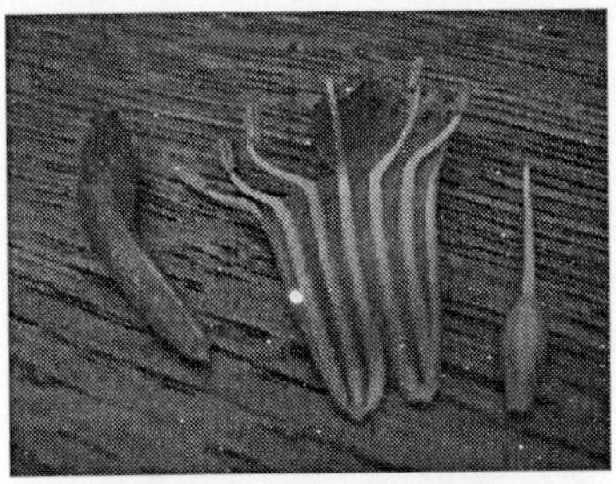

Fig.3.20: Epipetalous

2. **Epitepalous(Fig.3.20a):** Stamens adhere to perianth Eg. *Spider lilly.*

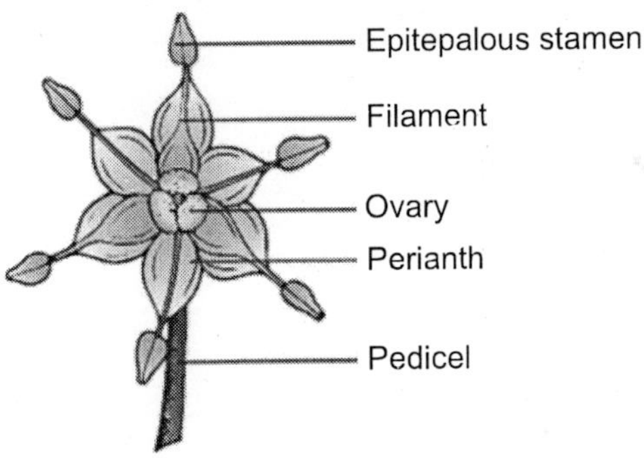

Fig.3.20a: Epitepalous

3. **Gynandrous(Fig.3.21):** Stamens adhere to the carpels Eg. *Calotropis.*

Gynandrous

Fig.3.21: Gynandrous

Based on length of stamens (Fig. 3.22)

1. **Didynamous:** Out of four stamens, two are long and two are short Eg. *Sesame.*
2. **Tetradynamous:** Out of six stamens, two outer stamens are shx stamens, two outer stamens are shorter and four inner stamens are long Eg. *Mustard.*

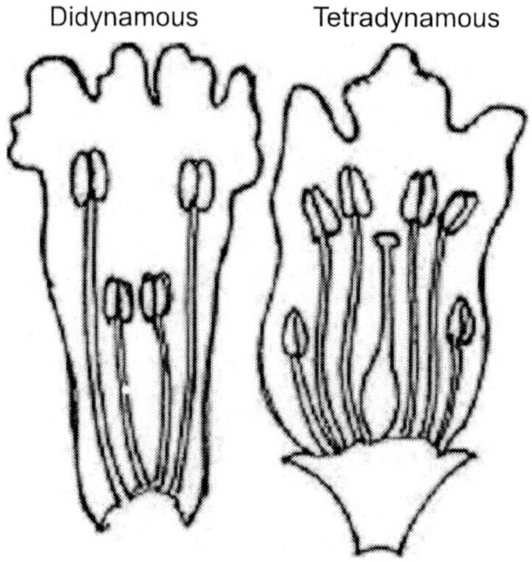

Fig. 3.22: Based on length of stamens

Based on fixation of anthers (Fig.3.23)

1. **Basifixed:** Filaments attached to base of anthers Eg. *Brassica.*
2. **Adnate:** Filaments continued from base to apex of anther Eg. *Verbena.*
3. **Dorsifixed:** Filament attached to dorsal side of anther Eg. *Citrus.*
4. **Versatile:** Tip of filament attached lightly at back side of anther. Hence, it can swing freely Eg. *Grasses.*

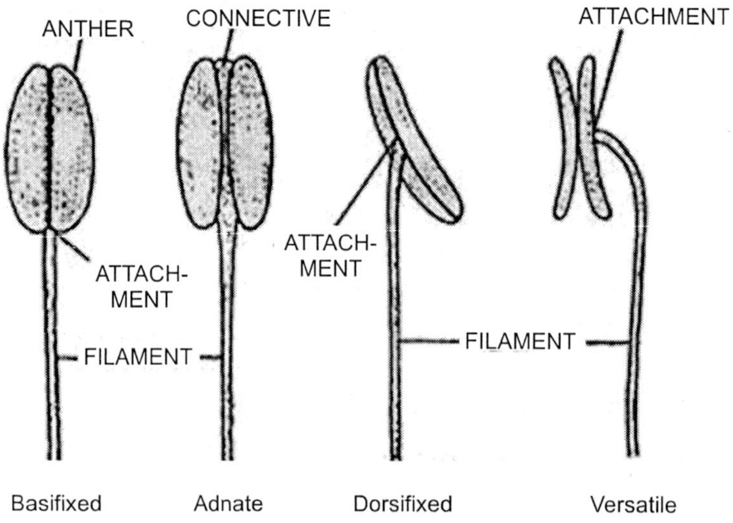

Fig. 3.23: Based on fixation of anthers

Gynoecium / Pistil: Innermost central whorl of flower which is female part of flower. A unit of gynoecium is called 'carpel (Megasporophyll)'. If a flower has a sterile ovules it is called as pistillode.

Based on number of carpels Gynoecium may be(Fig.3.24)

1. **Monocarpellary:** Gynoecium has single carpel Eg. *Fabaceae.*
2. **Bicarpellary:** Gynoecium has two carpels Eg. *Rubiaceae.*
3. **Tricarpallary:** Gynoecium has three carpels Eg. *Liliaceae.*
4. **Tetracarpellary:** Gynoecium has four carpels Eg. *Melia.*
5. **Multicarpellary:** Gynoecium has many carpels Eg. *Papaver.*

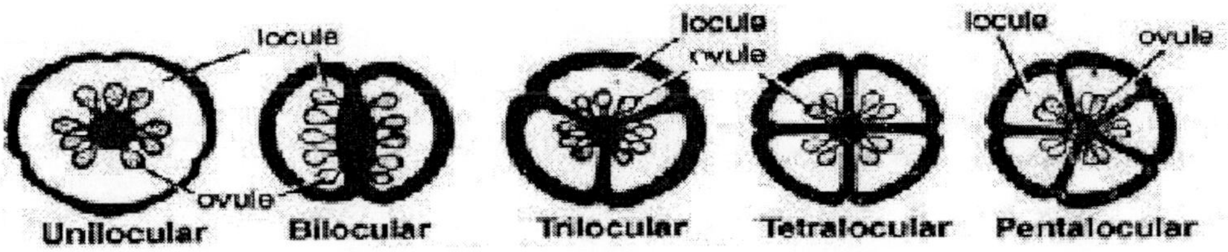

Monocarpellary Bicarpellary Tricarpallary Tetracarpellary Pentacarpillary

Fig. 3.24: Number of carpels Gynoecium

Based on cohesion of carpels (Fig.3.25)

1. **Apocarpous:** Gynoecium has two or more free carpels Eg. *Polyalthia.*
2. **Syncarpous:** Gynoecium has two or more fused carpels Eg. *Hibiscus.*

Apocarpus [with free carpels]

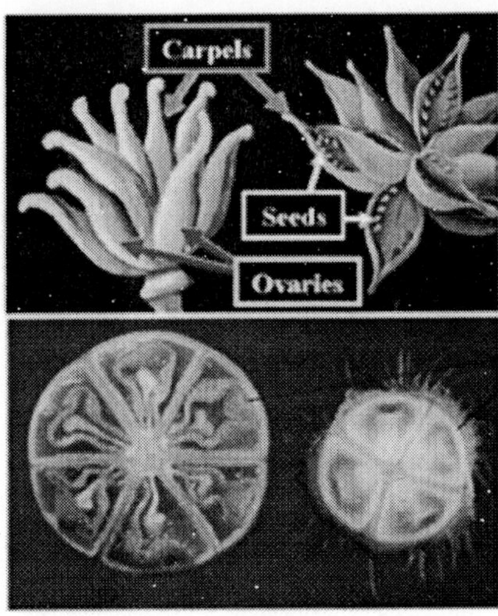

carpels

Syncarpous [with connate carpels]

Fig. 3.25: Based on cohesion of carpels

PLACENTATION

Ovules are present inside the ovary. Ovules are connected to the ovary with special type of tissue called placenta (Fig. 3.26). The mode of distribution of placenta inside the ovary is called placentation.

1. **Axile placentation:** Carpel walls meet in the centre of ovary where placenta are formed like central column. Eg. *Hibiscus esculentus* (*Bhindi*)
2. **Marginalplcentation:** Ovulesareborne along the junction of two margins of carpels Eg. *Fabaceae*.
3. **Parietal placentation:** Carpels are fused at which placenta has ovules Eg. *Cucumber*.
4. **Basal placentation:** Placenta develop directly on the receptacle which bears single ovule at base of ovary. Eg. *Asteraceae*.
5. **Superficial placentation:** Ovules are borne on placenta which develop all around the inner surface of the partition wall. Eg. *Nymphaeaceae*.

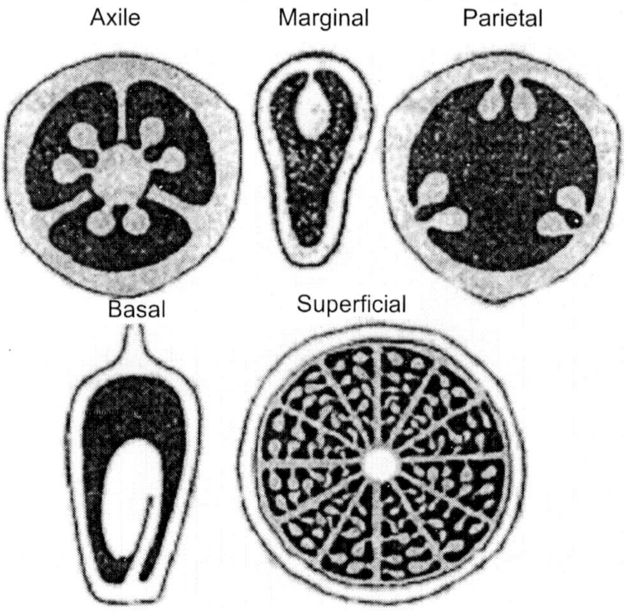

Fig. 3.26: Placentation

Based on floral whorls

- **Complete flower (Fig. 3.27):** All the four whorls are present in flower.

Parts of a flower

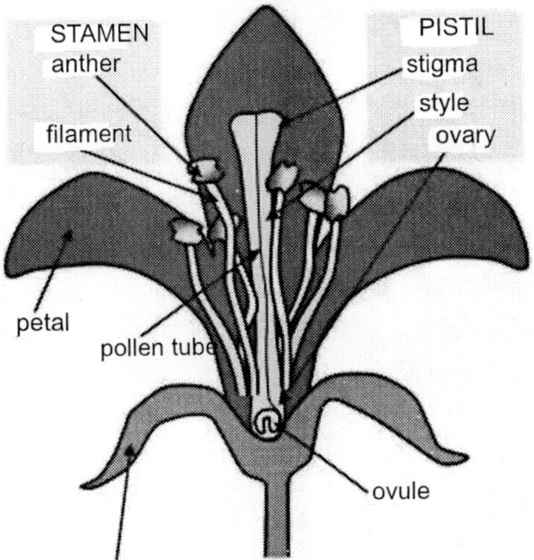

Fig.3.27: Complete flower

- **Incomplete flower(Fig.3.28):** One or more whorls are absent.

Fig. 3.28: Incomplete Flower

- **Regular flower(Fig.3.29):** The parts of each whorl are uniform in size and shape.

Fig. 3.29: Regular Flower

- **Irregular flower(Fig.3.30):** The parts of a whorl are unequal in size.

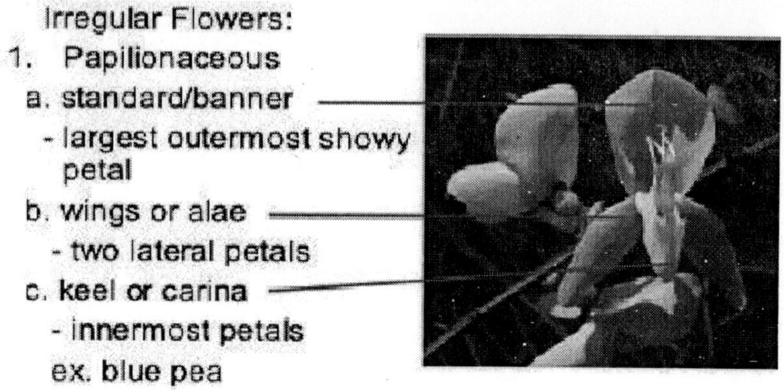

Fig.3.30: Irregular Flower

Based on accessory whorl distribution (Fig.3.31)

- **Monochlamydeous:** Only one accessory whorl (either calyx or corolla) is present.
- **Dichlamydeous:** Both calyx and corolla are present.
- **Achlamydeous:** Both calyx and corolla are absent and hence, flowers are naked.

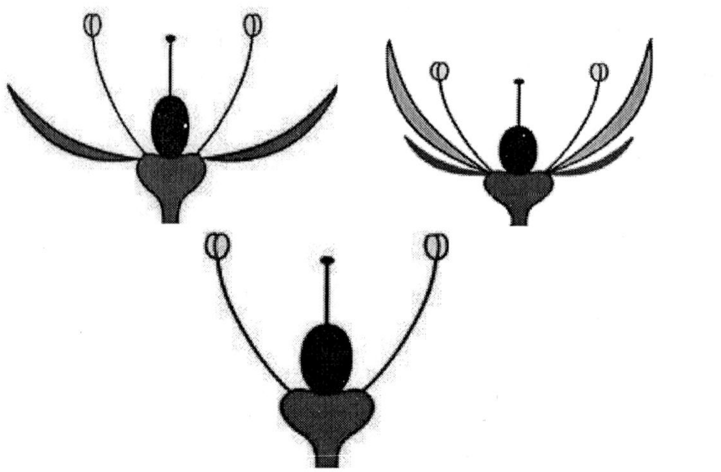

Monochlamydeous Dichlamydeous Achlamydeous

Fig. 3.31: Accessory whorl distribution

Based on sex distribution

1. **Bisexual/Perfect flower (Fig.3.32):** Both essential whorls viz., androecium and gynoecium are present.

Fig.3.22. Bisexual / Perfect flower

2. **Unisexual / Imperfect flower(Figs.3.33a,b,c,d,e)):** Only one essential whorl is present.
   (a) **Staminate(Fig.3.33a):** Male flower with androecium only.

Fig.3.33a: Staminate

   (b) **Pistillate (Fig.3.33b):** Female flower with gynoecium only.

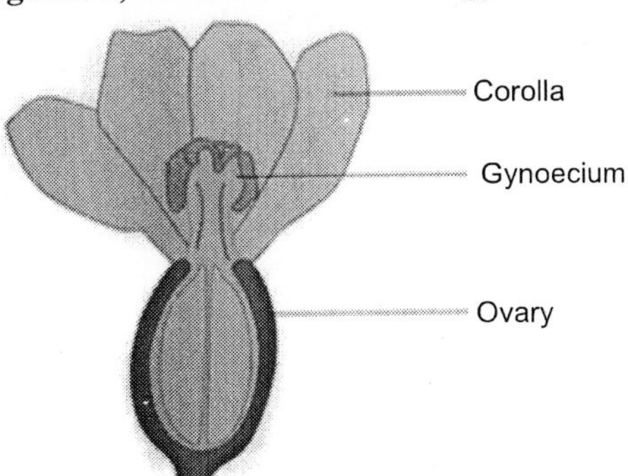

Fig. 3.33b: Pistillate

- **Monoecious (Fig.3.33c):** Male and female flowers occur on the same plant Eg. *Maize, Coconut.*

Fig.3.33c: **Monoecious**

- **Dioecious (Fig.3.33d):** Male and female flowers occur on separate plant Eg. *Papaya, Palm.*

Fig.3.33d: **Dioecious**

- **Polygamous (Fig.3.33e):** Plant has three kinds of flowers viz., staminate, pistillate and bisexual flowers Eg. *Mango, Cashew.*

Fig.3.33e: Polygamous

Floral symmetry

The shape, size and arrangement of floral whorls around the floral axis is called floral symmetry (Fig. 3.34).

1. **Actinomorphic:** A flower can be divided into two equal halves along more than one median longitudinal plane Eg. *Hibiscus*.
2. **Zygomorphic:**Aflowercanbedividedintotwoequalhalvesinonlyoneverticalplane Eg. *Pisum*.
3. **Asymmetric:** A flower can not be divided into two equal halves along any vertical plane Eg. *Canna*.

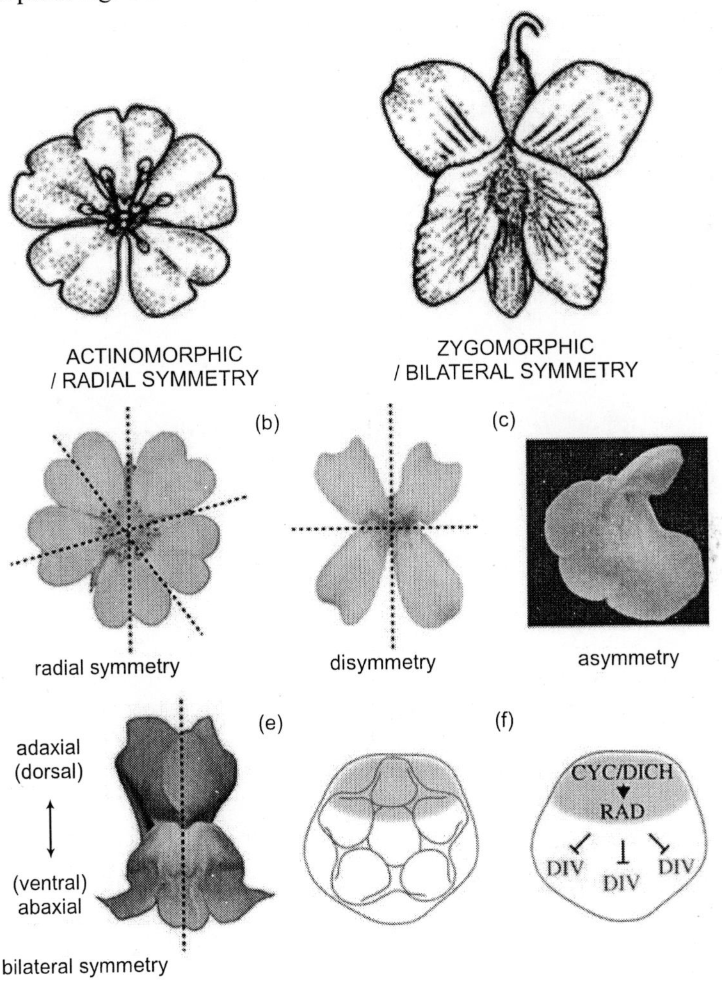

Fig. 3.35: Floral symmetry

Based on number of floral parts

Occurrence of same number of floral parts in different floral whorls of a flower is called isomery or isomerous flowers (Fig.3.36). Sometimes, flowers have different number of parts in each whorl is called heteromerous. The isomerous flower is of different types

- **Dimerous:** Floral parts are in two's or multiples of two.
- **Trimerous:** Floral parts are in three's or multiples of three.
- **Tetramerous:** Floral parts are in four's or multiples of four.
- **Pentamerous:** Floral parts are in five's or multiples of five.

Dicotyledonous flowers are usually tetra or pentamerous whereas monocotyledonous flowers are trimerous.

Fig. 3.36: Number of floral parts

FLORAL FORMULA

Floral formula is an expression summarizing the information in floral diagram by using symbols. Important symbols are following,

		Bracts and Epicalyx	A_5	-	Five free stamens
Br	-	Bracteate	A_{5+5}	-	Ten stamens in whorl of five each
Ebr	-	Ebracteate	A_0	-	Stamens absent
Brl	-	Bracteolate	$A\alpha$	-	Stamens indefinite in number
Ebrl	-	Ebracteolate	$\widehat{PA}$	-	Epiphyllous
E	-	Epicalyx	$\widehat{GA}$	-	Epipetalous
		Symmentry			**Gynoecium**
⊕	-	Actinomorphic	G	-	Gynoecium
%	-	Zygomorphic	$\underline{G}$	-	Superior ovary

		Calyx	$\bar{G}$	-	Inferior ovary
K	-	Calyx	G-	-	Half superior
K4	-	Four free sepals	G_2	-	Two free carpels
$K_{(4)}$	-	Four fused sepals	$G_{(2)}$	-	Two fused carpels
		Corolla	G_0	-	Carpels absent
C	-	Corolla	$\underline{G}_{(2)}$	-	Bicarpellary, superior ovary
C_4	-	Four free petals	$\underline{G}_{(2)}$ -	-	Bicarpellary, semi-inferior ovary
$C_{(4)}$	-	Four fused petals	$\bar{G}_{(2)}$	-	Bicarpellary, inferior ovary
		Sex			Perianth
⚥	-	Bisexual	P	-	Perianth
♂	-	Staminate	P_6	-	Six free tepals
♀	-	Pistillate	$P_{(6)}$	-	Six fused tepals
		Androecium	P_{3+3}	-	Six tepals in whorl of three each
A	-	Androecium			
$A_{(5)}$	-	Five fused stamens	∞	-	Indefinite

FRUIT

After fertilization ovary develops into fruit. Fruit is defined as fertilized and developed ovary. Fertilization acts as a stimulus for development of ovary. Sometimes, ovary may develop into fruit without fertilization called parthenocarpy and such fruits are called parthenocarpic fruits which are seedless. Eg. Banana, Grapes, Guava and Pineapple.

Structure of fruit

Fruit contains two main parts:

1. **Pericarp:** Ovary wall develops into fruit wall called pericarp.
2. **Seed:** Ovules inside ovary develop into seeds.

Pericarp consists of three layers:

- **Epicarp:** Outer layer.
- **Mesocarp:** Middle layer which is sweet, juicy and edible flesh.
- **Endocarp:** Inner layer.

Classification of fruit:

1. Simple fruit.
2. Aggregate fruit.
3. Multiple/Composite fruit.

1. Simple fruit

When a single fruit develops from single ovary of single flower, it is called as simple fruit. The ovary may be monocarpellary or multicarpellary syncarpous.

There are two types,

- Fleshy fruit
- Dry fruit

(i) Simple fleshy fruit(Fig.3.37):

Pericarp may be differentiated into an epicarp, a middle mesocarp and an inner endocarp. Entire or part of pericarp is succulent and juicy on ripening.

a) Drupe :
- One seeded fruit
- Pericarp is differentiated into
    - Skinny epicarp
    - Fleshy and juicy mesocarp
    - Hard and stony endocarp which encloses a single seed
- Edible portion is mesocarp. Eg. Mango
- In coconut, mesocarp is fibrous and endosperm is edible

b) Berry :
- Multi seeded fruit
- Epicarp is thin
- Meso and endocarp is undifferentiated to form a pulp in which seeds are embedded
- All parts are edible, Eg. Tomato

c) Hesperidium :
- Pericarp is differentiated into
    - Glandular epicarp
    - Fibrous mesocarp
    - Membranous endocarp
- Fruit chambers are formed and seeds arise on axile plancentation and they are covered by juicy hairs which are edible. Eg. Citrus

d) Pepo :
- A large multiseeded fleshy fruit develops from a tricarpellary, syncarpous, unilocular, inferior ovary with parietal placentation.
- The placenta and inner portion of pericarp become swollen and juicy which are edible. Eg. Cucumber, Melon, Bottle gourd

e) Pome :
- It is false fruit or pseudocarp
- Fruits develop from multicarpellary, syncarpous, inferior ovary containing seeds which remain inside
- Thalamus develops along with ovary and enclosing the fruit. Edible part is thalamus. Eg. Apple, Pear

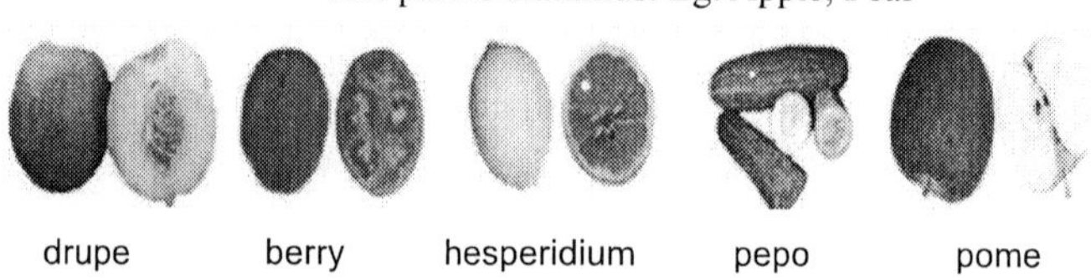

Fig. 3.37: Simple fleshy fruit

(ii) Simple Dry fruit(Fig.3.38)

Fruits have dry pericarp and the pericarp is not distinguished into three layers. These are different types.

(a) Dehiscent dry fruit: Fruits get dried up on maturity and splitted automatically to release seeds.

(b) Indehiscent dry fruit: Fruits get dried up on maturity but seeds are not released automatically. Only on decay of pericarp seeds are released.

(c) Schizocarpic dry fruit: Fruit dehisced up into number of indehiscent part (mericarp).

(a) Dehiscent dry fruit(Fig.3.38)

- Follicle
  - Dehiscent dry fruit develops from monocarpellary, superior ovary with marginal placentation
  - Fruit dehisces from only one suture. Eg. Calotropis
- Legume / pod
  - Dehiscent dry fruit develops from monocarpellary, superior ovary with marginal placentation
  - Fruit dehisces from both suture. Eg. Fabaceae
- Siliqua
  - Dehiscent dry fruit develops from bicarpellary, syncarpous, superior ovary with parietal placentation.
  - Ovary is unilocular but appears as bilocular due to false septum called replum to which large number of seeds are attached.
  - Fruit dehisces along both suture from base to apex. Eg. *Brassicaceae.*
- Silicula
  - Dehiscent dry fruit develops from syncarpous ovary having a false septum known as replum and it is short and flattened version of sliqua fruits.
- Capsule
  - Dehiscent dry fruit develops from bicarpellary to multicarpellary, syncarpous, superior ovary with axile placentation.
  - Fruit dehisces along 2 or more lines of suture. Eg. *Solanaceae, Malvaceae, Amaranthaceae.*

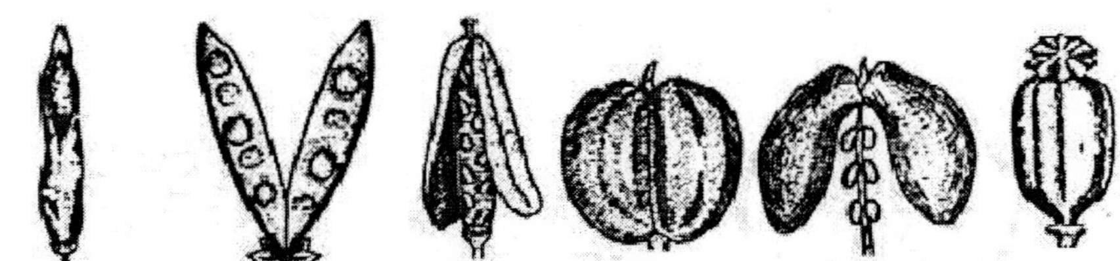

Follicle *(Delphinium)* Legume *(Pisum sativum)* Siliqua *(Mustard)* Silicula *(Iberis)* Silicula Dehisced *(Iberis)* Capsule *(Papaver)*

Fig.3.38. Dry and dehiscent fruits

(b) Indehiscent dry fruit(Fig.3.39)

- o Achene - One seeded fruit develops from a superior monocarpellary or bicarpellary syncarpous ovary with basal placentation.
  - Pericarp is hard and free from seed coat Eg. *Mirabilis.*
- o Achene - Feathery achene of clematis is achene of clematis.
- o Utricle - Small fruit with pericarp free from seed Eg. *Ragi.*

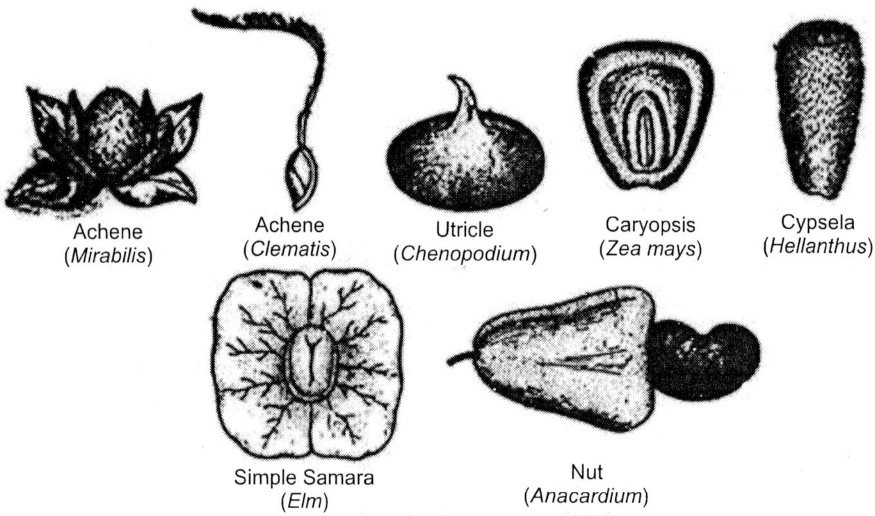

Fig.3.39: Indehiscent dry fruit

Caryopsis: A Caryopsis is a simple dry indehiscent fruit, like an achene, but with the seedcoat fused with the fruit coat Eg. *Zea mays* and all members of the Grass Family (*Poaceae*):

Cypsela: A Cypsela is a single-seeded dry indehiscent fruit that develops from a one part inferior ovary (on the salk side of the flower) Eg. Helianthus.

Simple samara: A simple samara is an independent dry indehiscent fruit which has part of the fruit wall extended to form a wing Eg. Elm (*Ulmus*).

Nut: A Nut is a large single hardened achene Eg. *Anacardium.*

(c) Schizocarpic dry fruit(Fig.3.40)

- o Lomentum - Modified legume having constriction between seeds
  - On maturity, fruit splits up into one seeded indehiscent compartments. Eg. *Tamarind*
- o Cremocarp - Two seeded fruit develops from inferior bicarpellary syncarpous ovary.
  - Two seeded fruit splits into two mericarp attached to carpophores Eg. *Coriandrum*

- Regma - Fruit develops from superior multicarpellary syncarpous with axile placentation
  - Fruit dehisces into many compartments called cocci Eg. *Castor, Geranium*

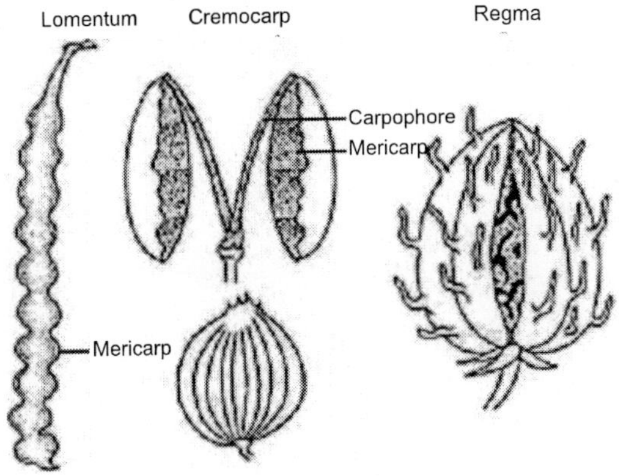

Fig.3.40: Schizocarpic dry fruit

2. Aggregate fruit(Fig.3.41)

Fruitlets develop from a single flower having multicarpellary apocarpous superior ovaries. Each ovary develops into simple fruitlets forming a bunch or etaerio of fruitlets. Such fruitlets remain separate from one another or coalesce together giving appearance as single fruit.

(i) **Etaerio of Follicles:** Each etaerio consists of two or more follicles, Eg. *Calotropis, Magnolia, Michelia.*

(ii) **Etaerio of Achenes:** Aggregate of achenes, Eg. *Clematis, Narvelia, Fragaria*

(iii) **Etaerio of Berries:** Several berries develop from multicarpellary apocarpous ovary embedded in fleshy thalamus Eg. *Polyalthia,* Custard apple.

(iv) **Etaerio of Drupes:** Bunch of several fruitlets aggregate on a fleshy thalamus Eg. *Rubus*

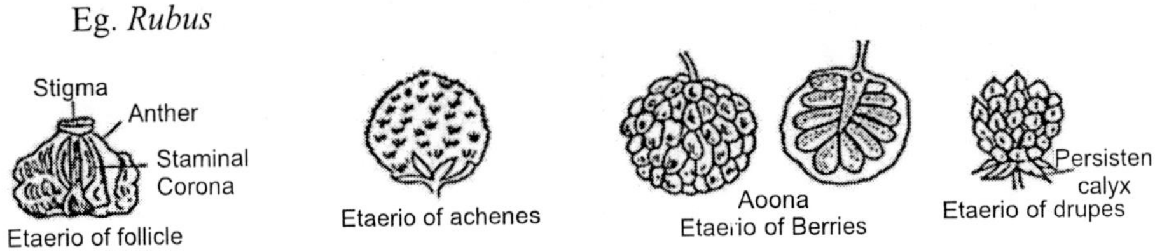

Fig. 3.41: Types of aggregate fruits

3. Multiple / Composite FruitLFig.3.42)

(i) **Sorosis:** The rachis and all floral parts of an inflorescence get fused together forming a composite fruit. Sessile flowers are arranged on fleshy rachis. Fertile fruits have juicy perianth and seeds covering membranous testa. Eg. *Jackfruit.*

Fig.3.42: Multiple / Composite Fruit

(ii) **Synconus(Fig.3.43):** Multiple fruit developed from hypanthodium inflorescence. The enlarged receptacle becomes fleshy and completely envelops the fruits. Eg. *Ficus*.

Fig. 3.43: Synconus

QUESTION BANKS

UNIT-I: SYSTEMS OF CLASSIFICATION AND GENERAL MORPHOLOGICAL DESCRIPTION

PART-A

0.5

MARK

CHOOSE THE BEST

1. Bentham and Hooker plant classification is based on which of following system — Key
   a) Artificial b) Natural c) Phylogenetic d) None of them — **b**
2. Which among the following plant is an example for tree fodder
   a) Danicha b) Sunhemp c) Agathi d) None of them — **c**
3. Which one of the crop is example for all the Narcotics, Fumitories and Masticatories
   a) Tobacco b) Opium c) Arecanut d) Betel leaf — **a**
4. Pulses seed are rich in
   a) Carbohydrate b) Protein c) Fat d) Fibre — **b**
5. Cereal crop grains are rich in
   a) Carbohydrate b) Protein c) Fat d) Fibre — **a**
6. An example for both green manures and fibre crop is
   a) Gliricidia b) Sunhemp c) Daincha d) Fibre — **b**
7. The binomial nomenclature includes
   a) Generic name b) Species name c) Race d) Both a & b — **d**
8. Binomial nomenclature was introduced by
   a) Bentham b) Hooker c) Linnaeus d) Darwin's — **c**
9. Sugar crop of tropical region is
   a) Sugarcane b) Sugar beet c) Sweet corn d) both a & b — **a**
10. Prop root is modification of
    a) Fibrous root b) Tuberous root c) Tap root d) None of them — **c**
11. Which of the following is an example of conical form root
    a) Radish b) Beet root c) Carrot d) Turmeric — **c**
12. The stem with hallow internode and solid node is called as
    a) Cladode b) Pilose c) Arborescent d) Culm — **d**
13. Which of the following is not a underground stem modification?
    a) Cladode b) Rhizome c) Corm d) Bulb — **a**
14. Which of the following is not an underground root modification? — Keys
    a) Napiform b) Fusiform c) Moniliform d) Stilt root — **d**
15. Which of the following is not an aerial stem modification?
    a) Cladode b) Creeping c) Twiners d) Runners — **d**
16. Which of the following is not a sub-aerial stem modification?
    a) Stolen b) Sucker c) Climbers d) Runners — **c**

17.	Which of the following is an example of palmately compound leaf					
	a) Tapioca	b) Mango	c) Cotton	d) Carrot	a
18.	The leaf of banana is					
	a) Palmately parallel	b) Palmately reticulate	c) Pinnately parallel	d) None of them	c
19.	The leaf of Castor is					
	a) Palmately parallel	b) Palmately reticulate	c) Pinnately parallel	d) None of them	b
19.	Arrow shaped leaf is referred as					
	a) Sagittate	b) Cordate	c) Lyrate	d) Retuse	a
20.	Which of the following is not a leaf modification?					
	a) Leaf tendrils	b) Phyllode	c) Leaf spines	d) Runners	d
21.	The mode of arrangement of leaves on the stem or branch is known as					
	a) Phyllotaxy	b) Aestivation	c) Venation	d) Both a & b	a
22.	Which of the following is an example of decompound leaf?					
	a) Tamarind	b) Carrot	c) Coriander	d) Both b & c	d
23.	Leaf with heart shaped leaf lamina is called as					
	a) Sagittate	b) Cordate	c) Lyrate	d) Retuse	b
24.	Which of the following is an example of cordate type of leaf lamina?					
	a) Betel vine	b) Thespesia	c) Both a & b	d) Neem	c
25.	Which of the following is not a palmately reticulate venation					
	a) Cotton	b) Cucurbits	c) Mango	d) Castor	c
26.	Inflorescence of rice is					
	a) Panicle	b) Spike	c) Arrow	d) Spadix	a
27.	Which of the following is an example of umbel Inflorescence?					
	a) Allium cepa	b) Oryza sativa	c) Zea mays	d) Piper nigrum	a
28.	Inflorescence of Sunflower is				**Keys**
	a) Spike	b) Capitulum	c) Corymb	d) Umbel	b
29.	Which of the following is not a mixed type of inflorescence?					
	a) Mixed panicle	b) Capitulum	c) Thyrus	d) Cymose umbel	b
30.	Which of the following is not a special type of inflorescence?					
	a) Cyanthium	b) Hypanthodium	c) Thyrus	d) Verticillaster	c
31.	When both male and female flowers are present in the same plant is called					
	a) Monoecious	b) Dioecious	c) Polygamous	d) None of them	a
32.	Which of the following is an example of Monoecious crop?					
	a) Mango	b) Papaya	c) Maize	d) Palmyra	c

33. A flower in which the ovary is superior and all other floral organs are situated below its level is called

    a) Hypogynous b) Epigynous c) Perigynous d) All of them **a**

34. A flower in which the ovary is inferior and other floral organs arise above it condition is

    a) Hypogynous b) Epigynous c) Perigynous d) All of them **b**

35. The expression of summarizing information in floral diagram by using symbols is

    a) Floral diagram b) Floral formula c) Floral structure d) All of them **b**

36. A kind of ground plan showing the arrangement of floral parts in a cross section of flower in bud condition is referred

    a) Floral diagram b) Floral formula c) Floral structure d) All of them **a**

37. Which of the following is not a fleshy fruit?

    a) Drupe b) Pome c) Berry d) Utricle **d**

38. Which of the following is not a dry fruit?

    a) Drupe b) Regma c) Caryopsis d) Utricle **a**

39. Fruit type of castor is

    a) Drupe b) Regma c) Caryopsis d) Utricle **b**

40. Fruit type of Rice is

    a) Drupe b) Regma c) Caryopsis d) Utricle **c**

41. The root system of Monocot is

    a) Tap root b) Fibrous root c) Both a & b b) Tuberos root **b**

42. International Botanical congress was formerly called as

    a) ICAR b) ICBN c) IBPGR d) BSI **b**

STATE TRUE OR FALSE

1. Bentham and Hooker plant classification is based on artificial system.
2. 202 families were recognized in Bentham and Hooker plant classification.
3. Botanically the characteristic fruit type of the family poaceae is a "caryopsis".
4. Cereals crops seed are rich protein.
5. Heroin drug is manufactured from opium plant.
6. Gliricidia and Azardiracta indica are example of green manures.
7. Daincha, and sunhemp crop is an example of green leaf manure.
8. Sunhemp is an example of green manure and fibre crop.
9. In general, the monocot plants leaves have reticulate venation.
10. The dicot plants leaves have parallel venation.
11. The leaf margin that is toothed is termed as serrated.
12. The tap root modification of carrot is conical form.

13. The tap root modification of beetroot is Fusiform.
14. The leaf of Castor is palmately parallel.
15. Stilt roots are occurred in maize and sugarcane.
16. Thyrus is a special type inflorescence.
17. Hypenthodium is a special type inflorescence.
18. A flower which is capable of being divided into two similar halves along any vertical plane is called actinomorphic flower.
19. A flower which is capable of being divided into two similar halves along only one particular vertical plane is called zygomorphic.
20. The expression of summarizing information in floral diagram by using symbols is called floral diagram.
21. Sunflower is an example for basal placentation.
22. Fruit type of rice is utricle.

Answer keys:

1. False 2. True 3. True 4. False 5. True 6. False 7. False
8. True 9. False 10. False 11. True 12. True 13. False 14. False
15. True 16. False 17. True 18. True 19. True 20. False 21. True
22. False.

FILL IN THE BLANKS

1. Bentham and Hooker system of classification was published in -------------- three volumes.
2. The organization that approves the bionomial nomenclature of plants is -----------
3. Expansion of ICBN -----------------.
4. The leaf of Moringa is a -------------- type of compound leaf.
5. The leaf of rice plant is ------------------ venation.
6. The venation type of banana leaf is --------------.
7. The arrangement and distribution of veins and veinlets in leaf lamina is called ----.
8. The arrangement of flowers on the floral axis is called --------------.
9. When the flower parts of each whorl is equal in size and shape is called -----------.
10. When the flower parts of a whorl is unequal in size and shape is called------------.
11. An example of monoecious plant is ---------------.
12. An example of dioecious plant is ------------------.

13. Arrangement of placentae and ovules in the ovary is called --------------.
14. Hesperidium type of fleshy fruit occurs in -------------.

Answer keys

1. Genera plantarum
2. ICBN
3. International Code of Botanical Nomenclature
4. Decompound
5. Palmately parallel
6. Pinnately parallel
7. Venation,
8. Inflorescence
9. Regular flower
10. Irregular flower
11. Maize/ castor/ coconut
12. Papaya/Palmyra/Mulberry
13. Placentation,
4. Lemon.

MATCH THE FOLLOWING

1.	Bentham and Hooker classification	- (a) Green leaf manure
2.	Gliricidia	- (b) Natural system
3.	Dhaincha	- (c) Green manure
4.	Cereals	- (d) Rich in protein
5.	Pulses	- (e) Rich in carbohydrate

Answer key

1. (b) 2. (a) 3. (c) 4. (e) 5. (d)

1.	Fusiform	- (a) Beetroot
2.	Napiform	- (b) Carrot
3.	Sweet potato	- (c) Banyan
4.	Money plant	- (d) Climbing root
5.	Prop root	- (e) Tuberous root

Answer key

1. (b) 2. (a) 3. (e) 4. (d) 5. (c)

1.	Nodulated root	- (a) Ginger
2.	Contractile root	- (b) Potato
3.	Rhizome	- (c) Cynodon
4.	Tuber	- (d) Onion
5.	Runner	- (e) Pulses

Answer key

1. (e) 2. (d) 3. (a) 4. (b) 5. (c)

1.	Leaf spine	- (a) Pea
2.	Trifoliate	- (b) Bladder wort
3.	Leaf tendrils	- (c) Cotton
4.	Leaf bladder	- (d) Black gram
5.	Palmately reticulate venation	- (e) Opuntia

Answer key

1. (e) 2. (d) 3. (a) 4. (b) 5. (c)

1.	Pome	- (a) Castor
2.	Balausta	- (b) Apple
3.	Regma	- (c) Redgram
4.	Utricle	- (d) Pomegranate
5.	Diadelphous	- (e) Ragi

Answer key:

1. (b) 2. (d) 3. (a) 4. (e) 5. (c)

PART-B

1 Mark

1. Differentiate monocots and dicots.
2. What is phyllotaxy? List out the types of phyllotaxy.
3. What is binomial nomenclature?
4. Define author citation.
5. Define aestivation and list out the types.
6. Write short notes on head or capitulum inflorescence.
7. What is hypogynous and epigynous?
8. Define floral diagram.
9. Define floral formula.

PART-C

2 Mark

1. Explain author citation and botanical nomenclature.
2. Outline the Bentham and Hooker system of classification.
3. Write merits and demerits of Bentham and Hooker system of plant classification.
4. What is Phyllotaxy? Explain types of phyllotaxy.
5. Define Venation. Explain types of venation.
6. Explain any two special type of inflorescence.

PART-D

5 Mark

1. Outline the Bentham and Hooker system of classification and write their merits and demerits.
2. Explain the agricultural classification of crops.
3. Elucidate the modification of stem.
4. What is racemose inflorescence and explain their types with suitable diagrams.
5. What is cymose inflorescence and explain their types with suitable diagrams.
6. Define aestivation. Elucidate types of aestivation with suitable diagrams.
7. Define fruit and explain their types.

Chapter 4

List of Cultivated Crops, Economic Parts, Chromosome Number, and Family description of Poaceae; Key Botanical features of Rice and Wheat

LIST OF CULTIVATED CROPS UNDER FAMILY POACEAE

Sl. No.	List of cultivated crops	Scientific name	Chromo-some number	Economic Parts
	Cereals			
1	Rice	*Oryza sativa*	24	Grain
2	Wheat	*Triticum aestivum*	42	Grain
3	Maize	*Zea mays*	20	Grain
	Millets			
4	Sorghum	*Sorghum bicolor*	20	Grain
5	Pearl millet	*Pennisetum glaucum*	14	Grain
	Minor / Small Millet			
6	Finger millet (Ragi)	*Eleusine coracana*	36	Grain
7	Foxtail Millet (Thenai)	*Setaria italica*	18	Grain
8	Little Millet (Samai)	*Panicum sumatrense*	36	Grain
9	Proso millet (Panivaragu)	*Panicum milliaceum*	36 & 72	Grain
10	Kodo millet (Varagu)	*Paspalum scrobiculatum*	40	Grain
11	Barnyard millet (Kuthirai valli)	*Echinochloa frumentosa*	36 & 54	Grain
	Forage grasses			
12	Guinea grass	*Panicum maximum*	18, 36,48	Whole plant
13	Napier grass (Elephant grass)	*Pennisetum purpureum*	28	Whole plant
14	Buffel grass	*Cenchrus setigerus*	32	Whole plant
	Sugar Crops			
15	Sugarcane	*Saccharum officinarum*	80	Stem

FAMILY DESCRIPTION OF POACEAE / GRAMINEAE

Plant	:	Annual / perennial herb
Stem	:	Hollow cylindrical internodes and swollen solid nodes
Leaf	:	Simple, Sessile, Sheathed leaf base, alternate, distichous, ligulate
Inflorescence	:	Panicle / Spike
Flower	:	Called spikelet consists of one or few florets with bract at the base called glumes, bisexual (Except maize which is monoecious), zygomorphic
Perianth	:	Represented by two lodicules
Androecium	:	Three stamens (Except rice which has six stamens)
Gynoecium	:	Monocarpellary, superior ovary
Fruit	:	Caryopsis (Except Ragi which has fruit type of Utricle)

SYSTEMATIC POSITION OF FAMILY POACEAE

Kingdom	:	Plant kingdom
Division	:	Phanerogams
Subdivision	:	Angiosperms
Class	:	Monocotyledon
Series	:	Glumaceae
Family	:	Poaceae

Reason for Prominence of cereals

- Greater adaptability.
- Easy for cultivation.
- Giving more yield per unit area due to tillering habit.
- Grains are dry and can be easily separated from plants.
- Grains have higher percentage of carbohydrate.

Pseudocereals:

Botanically unrelated plants whose seeds are used similarly as that of cereals. Eg. Buckwheat (*Agopyrum esculentum*), *Grain of Amaranthus*.

Botanical features of Rice

Rice – *Oryza sativa* $2n = 24$ Family: Poaceae

- It is principal crop of the world's population.
- It occupies third place in production after wheat and maize.

Habitat	:	Adapted to aquatic environment due to intercellular air space in leaves / root. Basically, tropical crop spread to subtropical and temperate region
Habit	:	Annual herb, Tillering habit
Root	:	Seminal root produced from radicle is later replaced by adventitious root which arise from basal node of the stem
Stem	:	Hollow cylindrical internodes, swollen and solid node, internode enclosed by leaf sheath, Internodal length is shorter at base and longer towards top, axillary bud at lower node develops into tillers
Leaves	:	Simple, sessile, sheathed leaf base, alternate, distichous, The junction of leaf sheath and leaf lamina called juncture / collar at which triangular membranous ligules (ligulate) and two sickle shaped auricle (auriculate) are present, linear leaf shape with parallel venation
Inflorescence	:	Treminal panicle, 14-42 cm long and has 50 – 500 number of spikelets. Panicle branched upto third order ended with spikelet. Ciliate ring present at the junction of culm and base of panicle. The distance between ciliate ring and top leaf determines the degree of exertion
Flower	:	Single pedicellate spikelet. zygomorphic, bisexual. It has bracts called glumes. Two lower glumes are sterile and two upper glumes are fertile called Lemma and Palea. Lemma is five nerved and palea is three nerved. Inside lemma and palea single floret is present.
Perianth	:	Represented by two lodicules, Polyphylious. Lodicules observe moisture and increase in size and pushing lemma and palea apart and thus helps in flower opening.
Androecium	:	Six stamens, slender filament, versatile anther, Polyandrous
Gynoecium	:	Monocarpellary, Superior ovary with two plumose stigma
Fruit	:	Caryopsis, The ovule after fertilization develops into the seed with its coats (outer cell layer) completely fused together with the developing ovary wall or pericarp. Under this circumstance, the fruit coat or the pericarp is fused with the seed coat.
Pollination	:	Self pollination

Floral Formula	:	Floral Formula= †,O, P(2), A(6), G1 Floral Formula: 6 stamens per Spikelet. 2 Lodicules are present (Presenting Calyx and corolla) forming Perianth. Protected by Lemma and Palea. Incomplete. Those flower which is missing one/two whorls. Perfect: Those flower which containing only essential whorls (Androecium & Gynoecium) Zygomorphic: Capable of division into symmetrical halves by only one longitudinal. Floral Formula = †,O, P(2), A(6), G1
Floral Diagram	:	As in fig 4.1

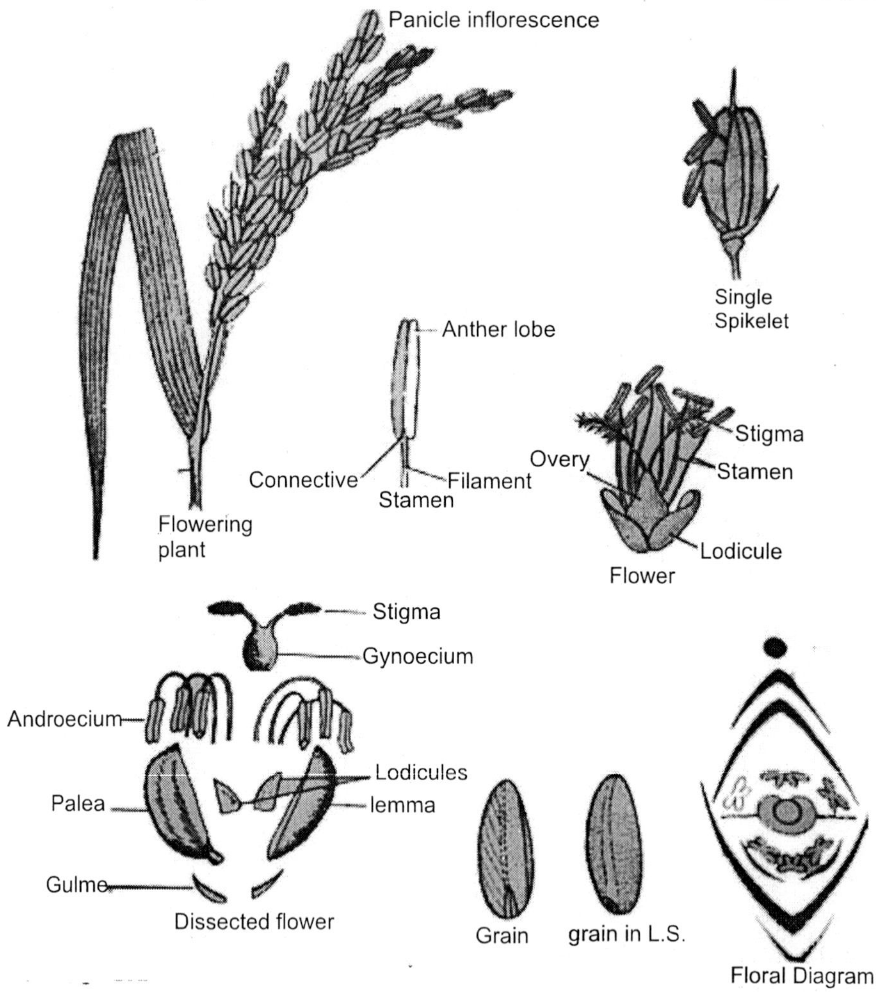

Fig. 4.1: Floral structure and floral diagram of *Oryza sativa*. L

Structure of Grain

Grain has following structures as discussed below:

1. **Pericarp or fruit coat:** Epicarp, mesocarp and endocarp are not distinguished. The colour in the rice grain is found in the pericarp layer in the mature stage.
2. **The seed coats:** Testa and tegmen pressed down below pericarp as integuments of seed coat.
3. **The aleurone layer:** A prominent layer of rectangular cells which contain protein lying next to the seed coats. This layer is known as the aleurone layer. This layer in rice is not coloured unlike in the case of maize.
4. **The endosperm:** The entire mass of tissue below the aleurone layer is made up of cells which contain plenty of starch grains and these form the endosperm.
5. **The embryo:** It contains plumule (embryonic leaf) and radicle (embryonic primary root) Plumule is enclosed by coleoptiles and radicle is covered by coleorhizae. These form embryonic axis which is bounded by scutellum.

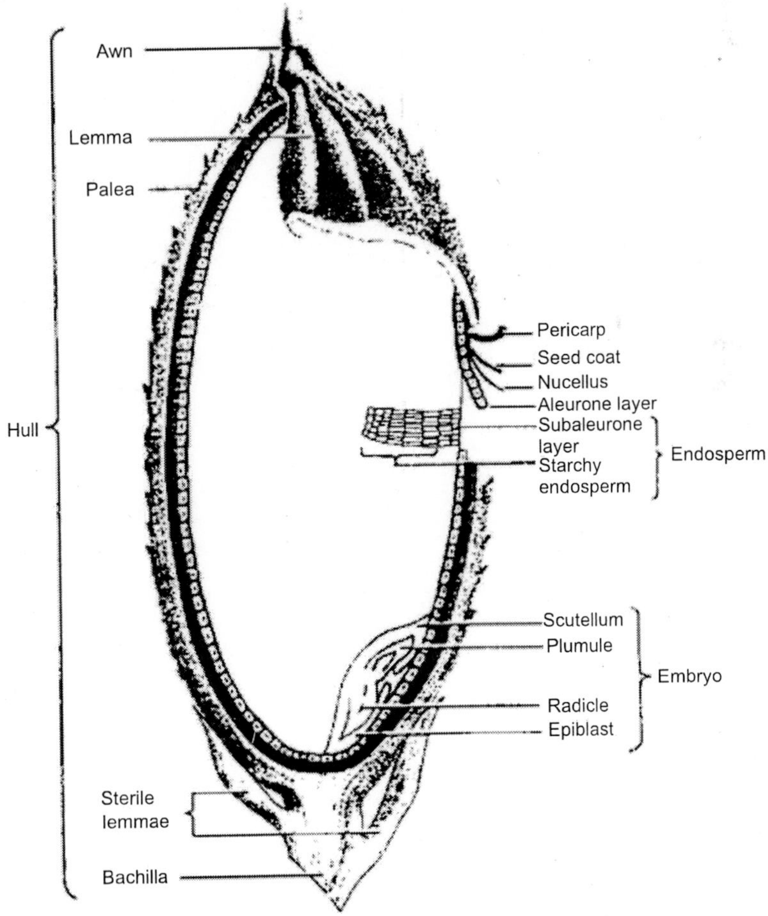

Fig. 4.2: Cross sectional structure of rice grain

Centre of origin: South East Asia

Cultivated species

1. Asian rice: *Oryza sativa* $2n = 24$ Origin: South East Asia
2. African rice: *Oryza glaberrima* $2n = 24$ Origin: West Africa

Putative parents: *Oryza sativa var. fatua*
Oryza perennis
Oryza breviligulata

Oryza sativa has the following three sub species

1. Tropical rice: *Oryza sativa sub sp. india*
2. Temperate rice: *Oryza sativa sub sp. japonica*
3. Java rice: *Oryza sativa sub sp. javanica*

Characters	Indica	Japonica	Javanica
Climate zone	Tropical	Temperate	Ewuatorial
Photoperiod	Sensitive	Insensitive	Insensitive
Duration	Long	Short	Very long
Response to manuring	Low	High	Low
Lodging	Susceptible	Non-lodging	Non-lodging
Grain shape	Long, narrow	Short, thick	Coarse, broad, thick
Plant colour	Light green	Dark green	Light green
Tiller number	Many	Moderate	Few
Grain shattering	Susceptible	Resistant	Resistant
Flag leaf	Long, narrow, drooping	Short, broad, erect	Long, wide semi drooping
Grain Breakage	High	Low	Low
Yield potential	Medium	High	Low
Market price	High	Low	Medium

Economic uses of rice

- Whole grain is cooked and eaten. Rice grain is used for producing parched rice, popped rice and beaten rice which are used as food.
- Rice flour is used for preparation of many edible item and is used in ice cream making.
- Starch from rice is used in textile industry and also for manufacture of dextrin and glucose.
- Rice bran is used as cattle feed.

- Rice bran oil is edible and also used for making soaps and cosmetic.
- Husk is used as fuel.
- Rice straw is used for cattle feeding.

Parboiling of rice

Paddy is steeped in water for 24 hours and then steamed under pressure. After steaming they are dried and milled.

Advantages of parboiling

- Preserves vitamins and minerals.
- Less breakage and high head rice recovery.
- Better keeping quality.

BOTANICAL FEATURES OF WHEAT

Wheat: Triticum aestivum $2n = 42$ **Family:** Poaceae

- It is most important staple food of world population.
- It is temperate crop grown mostly in Northern zone of India.
- It ranks second in production among cereals.

Habitat	:	It is temperate crop grown in varied types of soil
Habit	:	Annual herb, Tillering habit
Root	:	Seminal root produced from radicle is later replaced by adventitious root which arise from basal node of the stem
Stem	:	Hollow cylindrical internodes, Swollen and solid node, internode enclosed by leaf sheath, Internodal length is shorter at base and longer towards top, axillary bud at lower node develops into tillers
Leaves	:	Simple, Sessile, Sheathed leaf base, alternate, distichous, ligulate auriculate, Linear leaf shape with parallel venation
Inflorescence	:	Treminal Spike, Sessile spikelets are arranged in distichous manner on rachis
Flower	:	Single sessiloe spikelet. Zygomorphic, bisexual. Each spikelet has pair of sterile glumes. Inside glumes two to nine florets are present of which lower florets are perfect and upper florets are rudimentary. Each fertile floret has Lemma and Palea
Perianth	:	Represented by two lodicules, Polyphyllous
Androecium	:	Three stamens, slender filament, versatile anther, Polyandrous

Gynoecium	:	Monocarpellary, Superior ovary with two plumose stigma
Fruit	:	Caryopsis
Pollination	:	Self pollination
Floral formula	:	As in fig.4.3

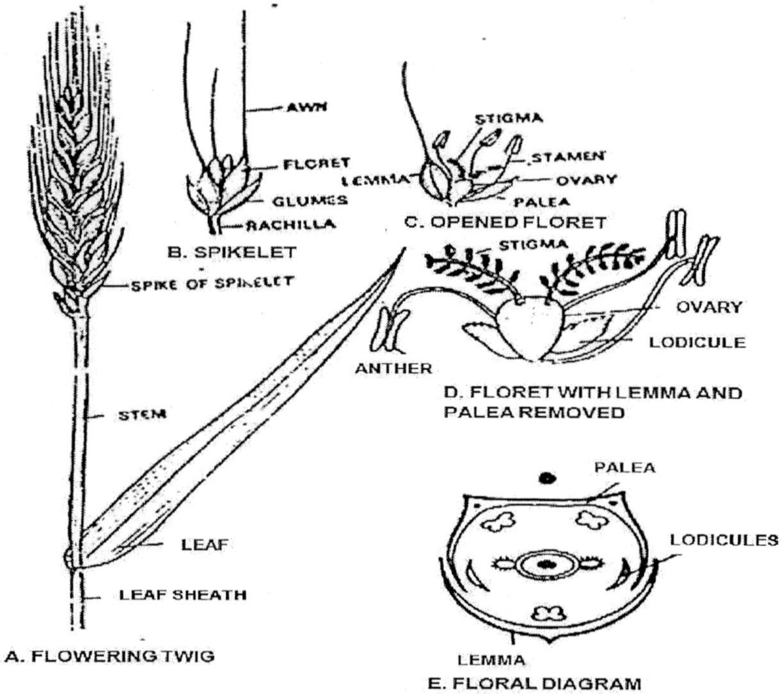

Floral formula: $\circ|\circ$ ☿ $P_{2\,(\text{lodicules})}\ A_3\ G_{\underline{1}}$

Fig. 4.3. Floral structure and floral diagram of Triticum aestivum

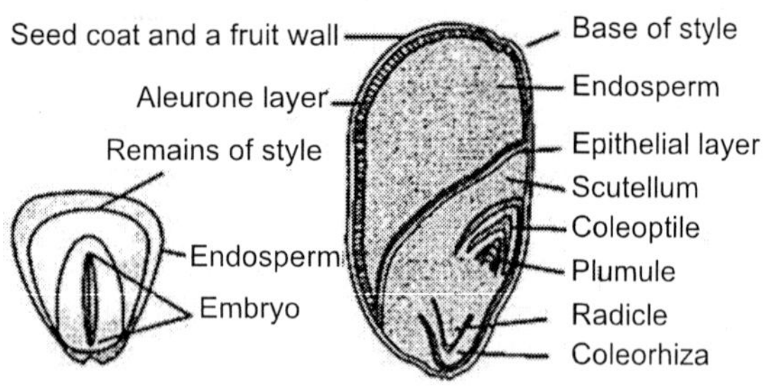

Fig. 4.4.Cross sectional structure of wheat grain

Centre of Origin:

- Diploid wheat : Asia minor
- Tetraploid wheat : Abyssinia and North Africa
- Hexaploid wheat : Central Asia

Origin of cultivated wheat

Triticum boeoticum x *Aegilops speltoides*
2n = 14 (AA) 2n = 14 (BB)

↓

F1
2n = 14 (AB)

↓ Chromosome doubling

Triticum dicoccoides x *Aegilops squarrosa*
2n = 28 (AABB) 2n = 14 (DD)

↓

F1
2n = 21 (ABD)

↓ Chromosome doubling

Triticum aestivum
2n = 42 (AABBDD)

Classification of Wheat

Based on season

1. Spring wheat: It is short duration wheat and does not require low temperature during initial growth stage.
2. Winter wheat: It is long duration wheat and require low temperature during initial growth stage.

Based on chemical composition

1. **Soft wheat:** It has low gluten content, pale coloured and posses white starchy area.
2. **Hard wheat:** It has high gluten content; dark coloured and has no white starchy area. Due to high gluten content hard wheat is suitable for bread making.

Economic Importance:

- Wheat is the staple food for most of the temperate and sub tropical regions of the world.
- The chief use of wheat is the flour for making bread biscuits, cookies, chapatti, etc.
- Industrially, it is used in preparation of starch, gluten, malt, distilled spirit.
- Wheat bran is rich in protein and used as valuable live stock feed.

Chapter

5

Key Botanical features of Maize, Sorghum, Pearlmillet and Finger Millet

BOTANICAL FEATURES OF MAIZE

Maize: Zea mays $2n = 20$ **Family:** Poaceae

- It is most important cereal after wheat and rice
- It is grown in wide range of climatic conditions

Habitat	:	It is grown in wide range of soils and climate
Habit	:	Stout annual grass, usually single stem
Root	:	Seminal root produced from radicle is later replaced by adventitious root which arise from basal node of the stem. During stem elongation 2 to 3 nodes above ground produce stilt roots which gives mechanical support to the plant and it is partially photosynthetic.
Stem	:	Thick, hollow cylindrical internodes, solid node, internode enclosed by leaf sheath, internodal length is shorter & thicker at base and longer towards top
Leaves	:	Simple, sessile, sheathed leaf base which encloses internode, alternate, distichous, ligulate, auriculate (auricle poorly developed), Linear leaf shape with parallel venation
Inflorescence	:	Monoecious
		Male inflorescence: Terminal panicle called tassel, branched
Flower	:	Spikelets are arranged in pairs on axis. Both are similar in structure. But one is sessile another one is pedicellate. Both spikelet have outer glumes which has two florets in side. Each floret has lemma and palea. Zygomorphic, staminate.
Perianth	:	Represented by two lodicules, polyphyllous
Androecium **Floral formula &Diagram**	:	Three stamens, slender filament, versatile anther, polyandrous In the fig 5.1 As in fig 5.1

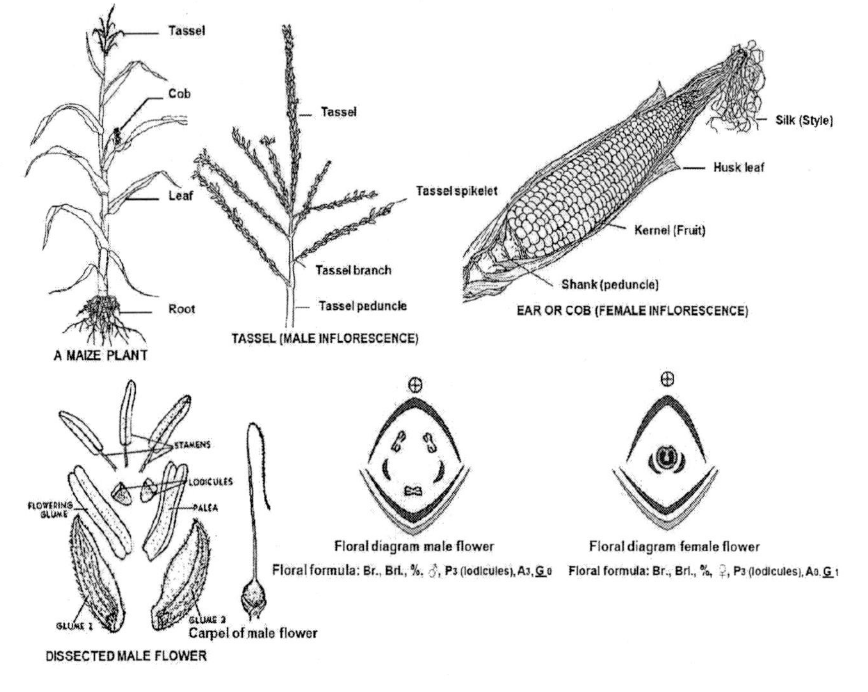

Fig. 5.1. Maize (Zea mays)

Female inflorescence	:	Axillary spike called Cob. Cob has reduced internode at base with modified leaves consisting of leaf sheath & reduced leaf blade called husk.
Flower	:	Sessile paired spikelets are arranged in longitudinal rows. Glumes, lemma and palea are reduced. Each paired spikelet is associated with a socket or cupule. Both the spikeles are sessile and identical. Each spikelet is two flowered, having a pair of small membranous glumes. The lower flower is non-functional, represented by a lemma and palea. The upper one is fertile and consists of a membraneous lemma and palea zygomorphic, pistillate
Gynoecium	:	Monocarpellary, superior ovary with long style. Bundle of style is called silk
Fruit	:	Caryopsis (kernel)
Pollination	:	Cross pollination due to protandry nature

Centre of Origin: Mexico

Putative parent: Introgression of Teosinte (Zea *Mexicana*) and Tripsacum (*Euchlana dactyloides*) played role in the evolution of maize.

Classification of Maize

Based on structure of grain:

1. **Pop corn (*Z. mays var. everta*):** The grains are small with a high proportion of very hard corneous endosperm with a little soft starch in the centre. On heating the steam generated inside the grain causes it to pop and explode, the endosperm becoming everted above the embryo and hull to produce a palatable white fluffy mass (popping).
2. **Flint corn (*Z. mays var. indurata*):** The grains consist of hard endosperm and a little soft starch in the centre. The strains are smaller than dent corn and have rounded ends.
3. **Dent corn (*Z. mays var. Indentata*):** The grain is characterized by a depression or dent at the top of the kernel having soft white starch surrounded by corneous endosperm on the sides. As the grain shrinks on drying produce a characteristic dent.
4. **Sweet corn (*Z. mays var. saccharata*):** The grains consist of a closely sweetish endosperm, which is translucent when immature, and dry to give a wrinkled appearance. It is popularly used as vegetable corn (Baby corn).
5. **Flour or soft corn (*Z. mays var. amylacea*):** The endosperm consists of soft starch and the grain usually has no dent.
6. **Waxy corn (*Z. mays var. ceratina*):** Endosperm waxy, which is composed of entirely amylopectin and used for manufacture of adhesives and also serves as an excellent substitute for tapioca flour.
7. **Pod corn (*Z. mays var. tunicata*):** It has no commercial importance but is of considerable interest in experimental studies concerned with breeding. It is a primitive type of maize.

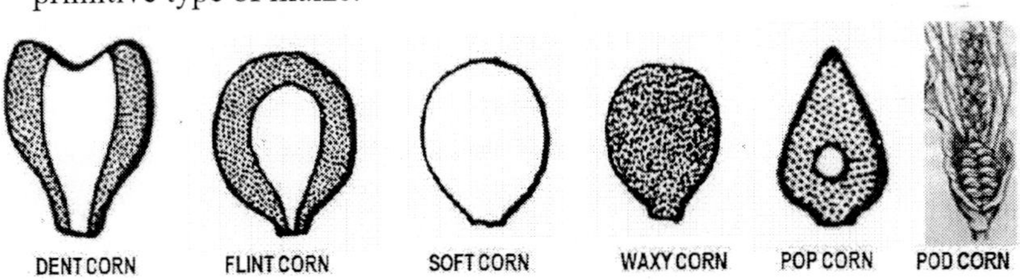

Fig. 5.2 Maize classified based on endosperm and grain structure

Economic Importance

As human food	:	Used as corn flake, corn meal, corn puff, popcorn, corn syrup, corn oil, corn rice.
As cattle feed	:	Grain as feed for poultry, piggery and cattle. Green fodder for silage and dry fodder.

Industrial uses

Corn starch	:	Textile
Corn grits	:	Corn flakes and starch
Brewer's grits	:	In breweries and distrilleries.
Corn maida	:	Bread, biscuit, vermicelli, bajji, pakoda etc. Also used in gum manufacturing industry.
Corn oil	:	Cosmetic, edible oil – improves blood circulation, reduce fat. etc., used as salad oil.
Corn syrup	:	Shoe polish and paper making
Corn sugar	:	Chemicals and leather industry
Glucose	:	Fermented liquor.
Zein	:	Utilised for making artificial fibers with good tensile strength and wool – like qualities.
Corn cake	:	Cattle and poultry feed.

Botanical features of Sorghum

Sorghum – *Sorghum bicolor* (L) Moench ($2n = 20$) **Family:** Poaceae

Sorghum is the fourth most important world food crop fllowing wheat, rice and Maize.

Habitat	:	Tropical and subtropical
Habit	:	Annual herbaceous, erect, single stalked or with tillers
Root	:	Seminal root produced from radicle is later replaced by adventitious root which arise from basal node of the stem. At lower nodes of stem stilt roots appear as a circle.
Stem	:	Erect, solid, 50 to 180 cm in height, slightly furrowed on alternate sides, pithy dry or juicy with light green thickened nodes. Internodes are short at the base and longer above.
Leaves	:	Simple, sessile, sheathed leaf base which encloses internode, alternate, distichous, midrib prominent, ligule is short, membranous and fringed present at the junction of leaf sheath and leaf blade. Below the leaf surface waxy coating will be there which prevents evaporation of water, Linear leaf shape with parallel venation.
Inflorescence	:	Usually compact panicle or semi compact or loose (lax) panicle. Terminal peduncle erect or recurved. Paired spikelets are arranged on rachis. One is sessile and other spikelet is pedicellate. Sessile spikelet is bisexual or hermaphrodite, Pedicelled one is male or sterile. Sessile fertile spikelet is comparatively larger than staminate spikelet. Sessile spikelet:

Flower	:	It has two glumes of approximately equal length (G_1 and G_2) having two florets inside; lower one is sterile with empty lemma (L_1) and no palea (P_1 absent); upper floret is perfect, bisexual consists of membraneous lemma (L_2) and palea (P_2), Zygomorphic.
Perianth	:	Represented by two lodicules, polyphyllous.
Androecium	:	Three stamens, slender filament, versatile anther, polyandrous.
Gynoecium	:	Monocarpellary, superior ovary, two plumose stigma.
Flower	:	Pedicellate spikelet, Spikelets are with long or short pedicel, two leathery boat shaped glumes enclose two florets. The lower floret is represented by the lemma (L_1) only and the upper floret is staminate with short awned lemma (L_2); palea (P_2) absent
Perianth	:	Represented by two lodicules, polyphyllous.
Androecium	:	Three stamens, versatile, polyandrous
Floral formula	:	In the fig 5.3
Floral diagram	:	As in the fig 5.3

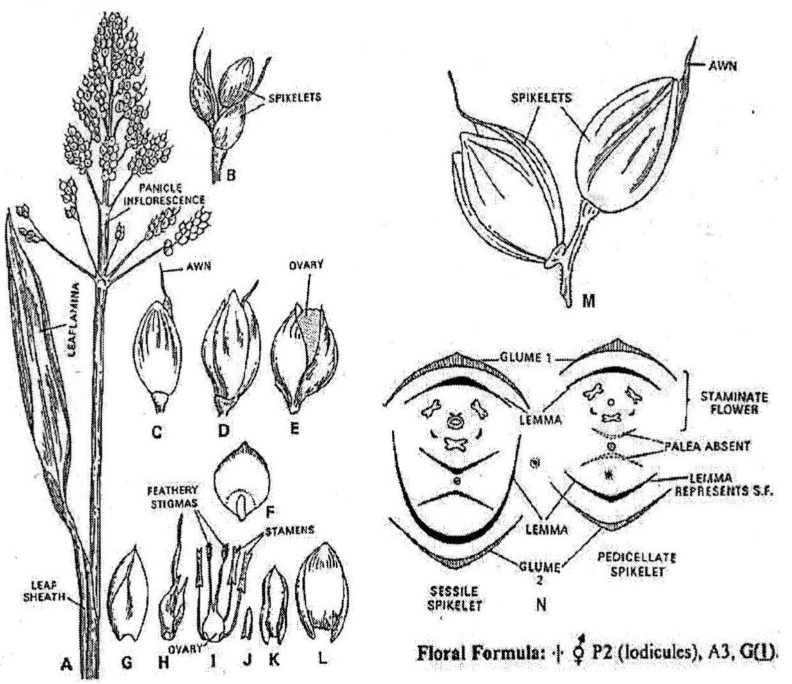

A, Panicle; B, Spikelet; C, Dorsal view of a spikelet; D, Ventral view; E. Opened caryopsis; F, A caryopsis; G, Glume; H, lemma; I, Ovary and stamens; J, Pale; K, Pale of fertile spikelet; L, Glume; M, Paired spikelets one sessile another pedicelled; N, Floral diagrams of sessile and pedicellate spikelet

Fig. 5.3: Floral structure and floral diagram of Sorghum bicolor

Fruit	:	Caryopsis (kernel)
Pollination	:	Often Cross pollination

Centre of Origin: Ethiopia

Putative parent: *Sorghum arundinaeceum*

Wild Relatives of Sorghum

The genus Sorghum is divided into six subgenera as follows:

1. **Eusorghum** – Cultivated, sorghum belongs Africa, India, Philippines
2. **Parasorghum** – South and Eastern Africa
3. **Chaetosorghum** – Australia
4. **Striposorghum** – Australia
5. **Sorghastrum** – Africa and America
6. **Heterosorghum** – New Guinea and Philippines, Australia

Classification of cultivated sorghum

The chief cultivated sorghum species classified following Snowden, 1936.

Section I: Eusorghum / Tree Sorghum

The Eusorghum is divided into two sub sections namely.

1. **Arundinaceae (2n:20):** Annual and tufted perennials without rhizome wild grasses and cultivated sorghum. This includes:
   (a) **Series:** Spontanea (Grass Sorghum); Includes S. sudanense, S. aethiopicum, S. virgatum, *S. verticifolium and S. stapfii.*
   (b) **Series:** Sativa (grain) : Includes S. vulgare, S.subglabresence and S. dochna.
2. **Halepenisia (2*n* : 20 & 40):** Perennial wild grasses with elongated rhizomes. includes *S. halepense, S. miliaceum, S. almum, S. propinquum and S. rodolphianum.*

The latest classification was done by Harlan and de Wet (1972)

1. **Bicolor (B):** Grain elongated, glumes, clasping the grain that may be completely covered or one fourth exposed. This includes:
   (a) *S. bicolor (Grain Sorghum)* – India Sweet in taste
   (b) *S. dochna (Broom corn)* – India and Burma, Principal broom sorghum
2. **Guinea (G):** Grains flattened dorsoventrally. This includes:
   (a) *S. guinense* (Guinea corn): Tropical W. Africa, Sudan and Uganda
   (b) *S. roxburghii* (Shallu): In East Africa, Indian and Myanmar; used as grain and fodder.
3. **Caudatum (C):** Grains asymmetrical, glumes half the length of the grain. This includes, *S. caudatum* (Feterita): Eastern and Western. Africa and Sudan; used as grain.

4. **Kaffir (K):** Grains symmedtrical, spherical, glumes clasping in varying length. This includes
   (a) *S. caffrorum* (Kaffir corn) – Principle crop in S. Africa.
   (b) *S. nigricans* (Beer Sorghum) – Equatorial Africa.
5. **Durra (D):** Grains rounded, obovate, wedge shaped at the base and broadest slightly in the middle; glumes very wide. This includes:
   (a) *S. drummondii* (chicken corn: Tropical W. Africa; grain not economical.
   (b) *S. cernuum* (White Durra): Asia Minor and India; used as grain and fodder.
   (c) *S. subglabrescens* (Milo): North East Africa and India; used as grain and fodder.
6. **Nervosum (N):** *S. nervosum* (Kaoling) – Central China; used as grain

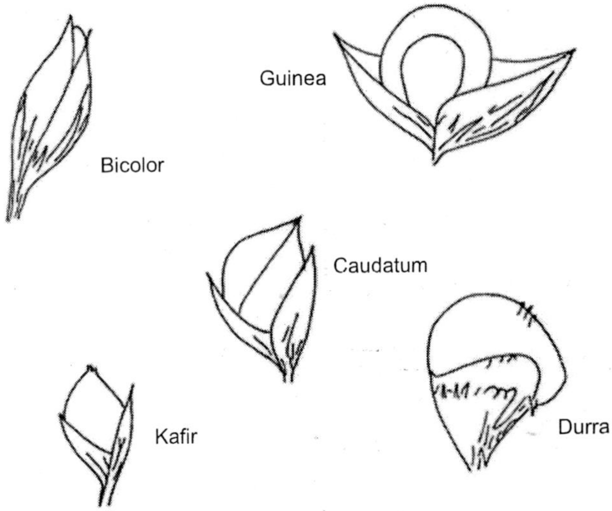

Fig. 5.4: Sorgum grain types based on species

Hybrid races

This consists of all combinations of the basic races namely.

1. Guinea bicolor (GB)
2. Caudate bicolor (CB)
3. Kaffir bicolor (KB)
4. Durra bicolor (DB)
5. Guinea caudatum (GC)
6. Guinea Kaffir (GK)
7. Guinea durra (GD)
8. Kaffir caudatum (KC)
9. Kaffir durra (KD)
10. Durra caudatum (DC)

Classification of Sorghum by Poehlman, 1959

1. **Grain Sorghum:** Principally, used for grain rather than fodder. Eg. Milo, Kaffir.
2. **Grass Sorghum:** It is used as fodder. Eg. Sudan grass (*Sorghum arundinaceum var. sudanese*), Tunis grass (*S. arundinaceum var. virgatum*) and Johnson grass (*S. halepense*).
3. **Special pupose Sorghum:** It includes cultivars with waxy endosperm which is used for starch manufacture and pop sorghum.
4. **Sorgos:** Principally used for fodder, hay, silage making and syrup production.
5. **Broom corn:** Inflorescence with long lateral branches used for making broom. Eg. *S. dochna.*

Wild Sorghum species of Tamil Nadu

1. *S. halepense and S.sudanense*: Utilized for forage sorghum improvement
2. *S. nitidum*: Found in Kodai Hills. Posses shoot fly resistance
3. *S. staffii*: Found in Southern districts; used for inducing dormancy

Land races of Sorghum in Tamil Nadu

1. Periamanjal cholam
2. Chinnamanjal cholam
3. Thalai virichan cholam
4. Makkattai cholam
5. Sen cholam
6. Irungu cholam
7. Vellai cholam

Economic Importance

- Flour of sorghum is used for making porridge, biscuits or unleavened bread.
- Pop sorghums with horny endosperm used for popcorn making.
- Sorghum is also widely used for brewing beer which is a valuable dietary supplement because of its high vitamin B content.
- Sorghum with sweet stems containing upto 10 per cent sucrose is used for brewing and manufacture of syrup.
- Sweet Sugar varieties containing 18 per cent total soluble sugar (TSS) is used for ethanol extracted.
- The grain is also used as valuable stock feed.
- Fodder sorghum principally used as fodder, hay and silage making.

Sorghum Poison

The shoot of sorghum contains cyanogenic glycoside called 'Dhurrin' which on hydrolysis by enzyme action gives rise to hydrocyanic acid (HCN). 0.5 g of HCN is sufficient to kill a cow. But, this poison is destroyed when the fodder is made into hay / silage.

BOTANICAL FEATURES OF PEARLMILLET

Pearl millet (Bajra, Cumbu): Pennisetum glaucum $2n = 14$ **Family:** Poaceae

- Staple food in drier parts of Africa and India
- It can be grown in poor sandy soil and in low rainfall areas

Habitat	:	Tropical and subtropical
Habit	:	Erect, annual, tillering habit and growing up to three meters.
Root	:	Seminal root produced from radicle is later replaced by adventitious root which arise from basal node of the stem.
Stem	:	Solid, slender or stout, round. Nodes slightly swollen with ring of silky while internodes are cylindrical and glabrous. The internodal length increases from the base of the culm upwards.
Leaves	:	Simple, Sessile, Sheathed leaf base which encloses internode, alternate, distichous, midrib prominent, The base of the blade is slightly auricled. The ligule is narrow membranous with fringe of hairs. Linear leaf shape with parallel venation
Inflorescence	:	Terminal contracted panicle or false spike, Cylindrical and tapering at apex. Cluster of one or two spikelets arranged spirally on the rachis. Spikelets subtended by bristles called involucres of bristles. Each spikelet has two florets enclosed in two glumes. The lower floret is staminate or sterile and the upper floret is bisexual. Lower floret:
Flower	:	Lower floret contains Lemma 1 and Palea 1 is absent, staminate zygomorphic
Perianth	:	Lodicules absent
Androecium	:	Stamens three with characteristic pennicillate anthers (hairs at the tip of the anthers), Polyandrous

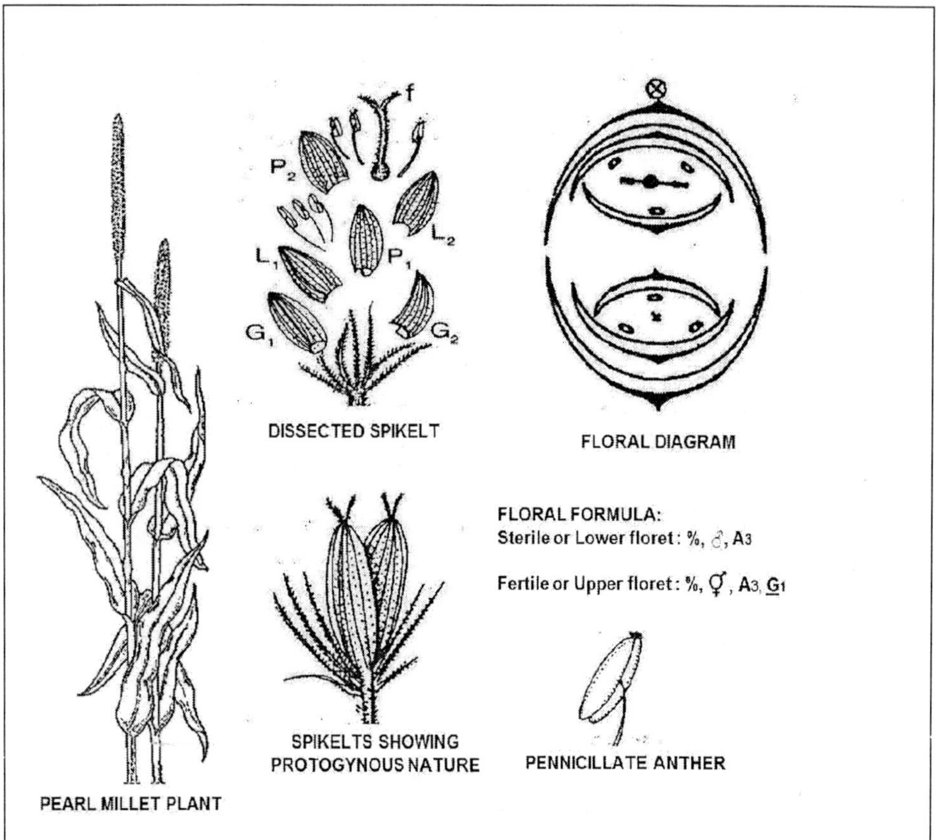

Fig. 5.5: Floral structure and floral diagram of Pearl millet

Gynoecium	:	Monocarpellary, superior ovary with bifid stigma
Fruit	:	Caryopsis. Well exerted from lemma and palea
Pollination	:	Cross pollination due to protogyny nature

Centre of origin: West tropical Africa

Putative parent: not known

Wild relatives:

Pearl millet are derived from two sources, one being the tall widely spread *P. purpureum* and the other the xerophytic group of smaller species of the section Penicillaria to which *P. glaucum* belongs. *P. purpureum* is a tetraploid (stapf and Hubbard, 1934). Other species include *P. spicatum* (India) *P. nigritarum* and *P. cinereum*.

Economic Importance

- **Flour :** Preparation of cakes or unleavened bread.
- **Grain :** Malted seed is an important source of beer, feed for poultry and other livestock.

- **Green plant :** as Fodder.
- **Straw :** Feed for livestock, bedding, thatching, fencing and fuel.

BOTANICAL FEATURES OF RAGI

Ragi (Finger millet) – Eleusine coracana $2n = 36$ **Family:** Poaceae

Habitat	:	Tropical and subtropical
Habit	:	Annual herb with tillering habit
Root	:	Seminal root produced from radicle is later replaced by adventitious root which arise from basal node of the stem.
Stem	:	Compressed, elliptic, bearing many distichous leaves.
Leaves	:	Simple, sessile, sheathed leaf base which encloses inter-node, alternate, distichous, ligule with fringe of hairs, lin-ear leaf shape with parallel venation
Inflorescence	:	Terminal digitate spikes, inflorescence borne on a long pe-duncle from the end of which four to five spikes radiate in a whorl called fingers with an odd one a little lower down the whorl called as the thumb. Rachis of the spikes are flat; spikelets are sessile, arranged in two rows alternately attached to one side of the rachis. Lower floret.
Flower	:	Each spikelet is having 3-7 florets enclosed by common glumes (G_1 and G_2). Florets are bisexual with lemma and palea, zygomorphic, terminal floret sterile
Perianth	:	Represented by two lodicules
Androecium	:	Stamens three, versatile anthers, polyandrous
Gynoecium	:	Monocarpellary, Superior ovary with two plumose stigma
Floral formula	:	As in thc fig. 5.6
Floral diagram	:	As in the fig . 5.6

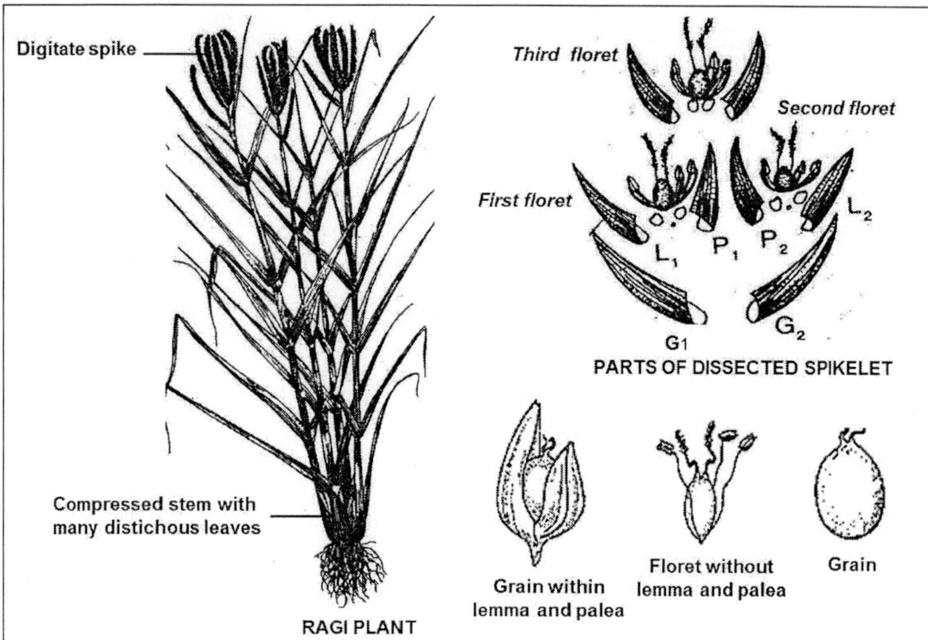

Fig. 5.6: Floral structures of Eleusine corocana

Fruit	:	Utricle
Pollination	:	Self pollination

Centre of Origin: Uganda

Putative parents: *Eleusine indica* $2n = 18$ Diploid

Eleusine africana $2n = 36$ Allopolyploid

Classification of cultivated ragi

The cultivars of E. coracana differ from the wild species *E. indica and E. africana* in having larger seeds and the spikelets do not shatter at maturity. Mehra (1963) recognize two groups of E. coracanae as noted below.

1. **African high land types:** Bearing resemblance to *E. africana*, with long spikelets, long glumes, and long lemmas with grain enclosed within the florets.
2. **Afro-Asiatic types:** With close resemblance to *E. indica, short spikelets, short glume, short lemmas,* and the mature grains exposed out of the florets.

Economic importance

- **Grain:** Food in the form of cake, porridge, sweet meat. Used for malting and brewing flour for making different edible things. Grains can be stored for long periods upto ten years or more without deterioration or weevil damage. It is an important famine food.

- **Ragi malt:** Food for the aged and children.
- **Straw:** Cattle feed.

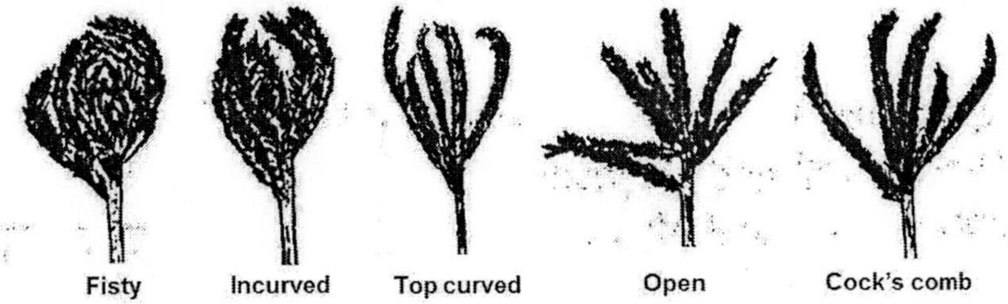

Fig. 5.7: Types of Ragi based on ear head shape

MINOR MILLETS

(7). FOX TAIL MILLET / ITALIAN MILLET/ THENAI- Setaria italica (2n:18) Family: Poaceae

Botanical Descroption

Annual grass; seminal roots three followed by numerous thin adventitious roots, culms erect, slender, internodes hollow, tillering; leaf sheath longer than internodes, ligulate; leaf blade linear; tip acuminate; mid rib prominent.

(a) **Inflorescence:** spike like panicle, carrying 6-12 two flowered sub sessile spikelets each subtended by 1-3 bristles; stamens three; ovary with two long styles ending in plumose stigma.

(b) **Fruit:** Caryopsis, tightly enclosed by lemma and palea.

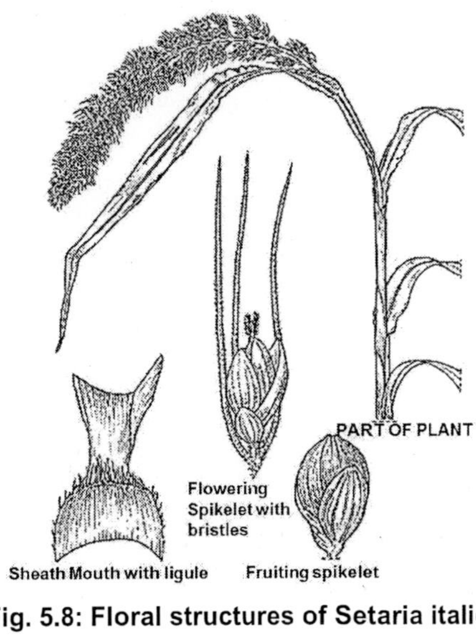

Fig. 5.8: Floral structures of Setaria italica

LITTLE MILLET (SAMAI) -*Panicum sumatrense* (P. miliare) (2n: 36)

Family: Poaceae

Botanical Description

An annual tufted grass with slender culms, soft leaves, inflorescence a panicle with erect hairy branches; spikelets in pairs with two glumes; floret with two lemmas, two lodicules, three stamens and ovary with plumose stigma; fruit a caryopsis.

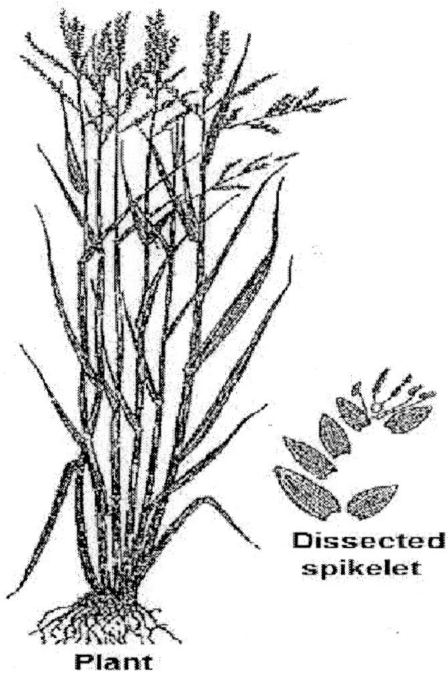

Fig. 5.9: Little millet: Panicum sumatranse (Poaceae).

PROSO MILLET (Panivaragu) -*Panicum milliaceum* (2n: 36 & 72)

Family: Poaceae

Botanical Description

A shallow rooted, erect annual grass, free tillering, internodes hollow, cylindrical; leaf lamina linear lanceolate. Inflorescence a slender panicle; spikelet with two florets with two glumes; lower floret sterile; upper floret fertile with lemma, palea, two lodicules, three stamens and two styles with plumose stigmas; fruit a caryopsis enclosed by persistent lemma and palea.

Fig. 5.10: Proso millet: Panicum milliaceum (Poaceae)

KODO MILLET (Varagu)-*Paspalum scrobiculatum* (2n: 40)

Family: Poaceae

Botanical Description

An annual tufted grass; leaves in two ranks, stiff, erect.

(a) Inflorescence: a panicle; 2-8 spikelets in flattened rachis; spikelets usually in two rows; each spikelet has two florets; lower floret sterile, upper bisexual with lemma, palea, two lodicules, three stamens and plumose stigma; grain enclosed in hard horny persistent husk which is difficult to remove.

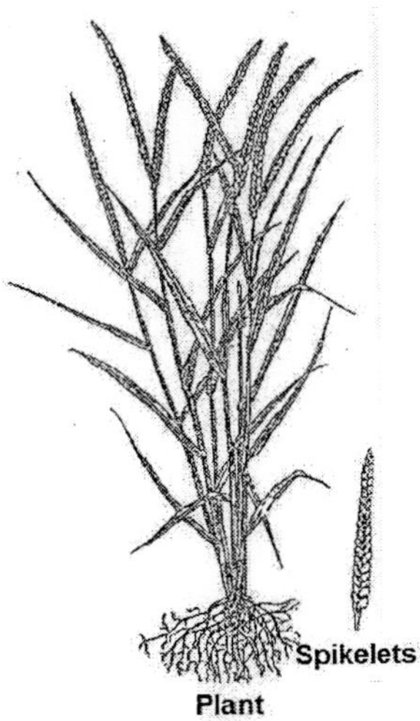

Fig. 5.11: Kodo millet: Paspalum scrobiculatum (Poaceae)

BARN YARD MILLET (*Kuduraivali*)

Echinochola frumentosa (2n: 36 and 54) Family: Poaceae

Botanical Description

A robust tufted annual grass; seminal roots followed by adventitious roots; stem smooth, glabrous, producing tillers; internodes hollow; leaf blade linear, lanceolate; tip acute; margin finely toothed. Inflorescence a panicle; spikelet two flowered, awnless, pedicellate, subtended by bristles, two glumes; lower floret sterile with lemma and palea; upper floret hermaphrodite, five nerved lemma and five nerved palea, two lodicules, three stamens, two distinct style with plumose stigma. Fruit a caryopsis enclosed in white shining hardened lemma and palea.

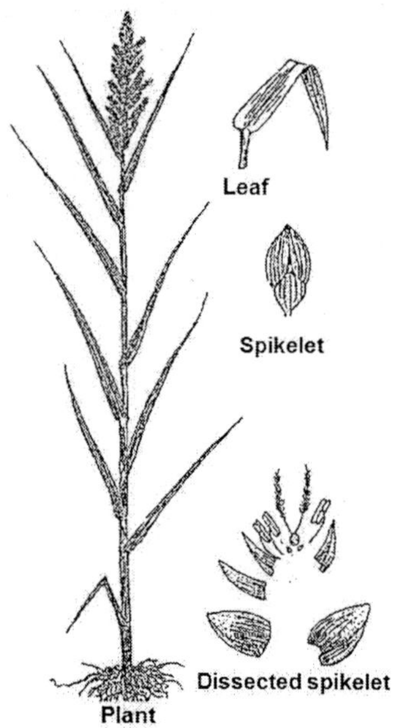

Fig. 5.12: Barnyard millet: Echinochola frumentosa (Poaceae).

Chapter

6

Key botanical features of Guinea grass, Napier grass, Cenchrus and Sugarcane

BOTANICAL FEATURES OF GUINEA GRASS

Guinea grass: *Panicum maximum* $2n = 18, 36$ & 48 **Family:** Poaceae

- It is forage grass grown along river bank and it is drought tolerant grass

Habitat	:	Tropical
Habit	:	A dense, erect, perennial, rhizomatous grass
Root	:	Seminal root produced from radicle is later replaced by fibrous root which arises from basal node of the stem. Shortly rhizomatous
Stem	:	Erect or geniculately ascending culms
Leaves	:	Simple, sessile, leaf blade long, upto 60 cm., and 2-2.5 cm broad, little rough and succulent, non droopy alternate, distichous, linear with parallel venation.
Inflorescence	:	Open terminal panicle with stiff branches in whorls. Spikelets small glumes two, Lemma and Palea, stamens three, ovary with two plumose stigmas
Floral formula	:	
Fruit	:	Caryopsis
Pollination	:	Often Cross pollination
Centre of Origin	:	Tropical Africa

Economic Importance

- It is a very good grass for grazing, green fodder and silage making grown in many sewage farms.
- It is free from all known toxic principles.
- It contains 5-8% protein with well balanced calcium and phosphorus.
- It is useful as a soil binder to protect against soil erosion.

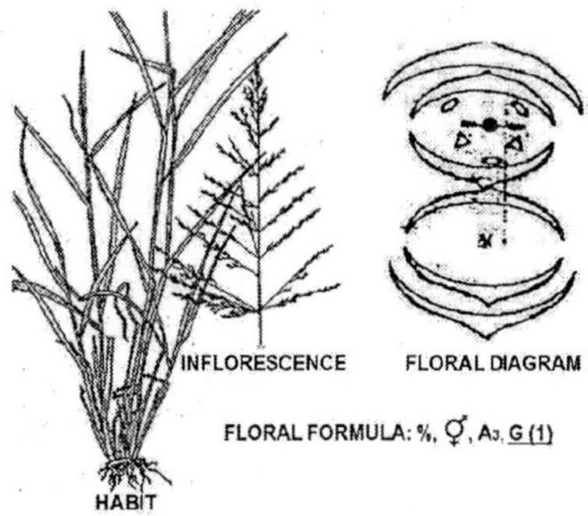

Fig. 6.1: Habit, floral diagram and structure of guinea grass (Megathyrus maximus)

BOTANICAL FEATURES OF NAPIER GRASS

Napier grass – *Pennisetum purpureum* $2n = 28$ **Family:** Poaceae

Habitat	:	Tropical
Habit	:	Perennial, tall, profuse tillering with creeping rhizomes.
Root	:	Fibrous root
Stem	:	Grows upto 7.5m tall, branched towards top
Leaves	:	Linear, long, dull green with hairy ligules.
Inflorescence	:	Terminal, dense, spiciform panicle. Yellow or tinged with brown or purple spikelets lanceolate, one or two flowered, G1 and G2 minute. Lemma one paleate, staminate lemma two paleate, bisexual, stamens three, penicillate anthers, style long, connate bifid stigma.
Floral formula	:	
Floral diagram :		

Fig. 6.2. Napier grass: Pennisetum purpureum (Poaceae)

Fruit	:	Caryopsis (kernel)
Pollination	:	Cross pollination and Apomictic
Centre of Origin	:	Tropical Africa

Economic Importance

- Leaf contains 5-7% protein with balanced proportion of calcium and phosphorus.
- Very palatable and quality forage
- Drought tolerant

BOTANICAL FEATURES OF CENCHRUS

Kolukattai grass: $2n = 32, 36, 40$ & 54 **Family:** Poaceae

White kolukattai : *Cenchrus ciliaris*

Neela Kolukattai : *Cenchrus glaucus*

Black klolukattai : *Cenchrus setigerus*

(Buffel grass)

Habitat	:	Tropical
Habit	:	Perennial grass with underground rhizomes.
Root	:	Fibrous root
Stem	:	Aerial branches, tufted, erect or decumbent, tillering.
Leaves	:	Leaf sheath slightly compressed, shorter than the internodes with scattered hairs, ligule short, thin, membrane fringed with hairs, the leaf blade is linear with hairs scattered above and sparse at the lower surface.
Inflorescence	:	A raceme of spikes, floral axis flattened and grooved, one to three spikelets subtended by involucres of bristles, connate at the base, spikelets sessile, oblong lanceolate, first glume is small, hyaline, ovate – lanceolate, second glume ovate, acute hyaline and one – third to half the length of the lemma, first lemma paleate, staminate or sessile, second lemma paleate with hermaphrodite flower, stamens three ovary with two plumose stigmas, lodicules absent, gain small, oblong.
Floral formula	:	As in Fig.6.3
Floral diagram	:	As in Fig.6.3
Fruit	:	Caryopsis. Well exerted from lemma and palea
Pollination	:	Cross pollination due to protogyny nature
Centre of origin	:	Tropical Africa

Economic importance

- It can be grown in drier areas

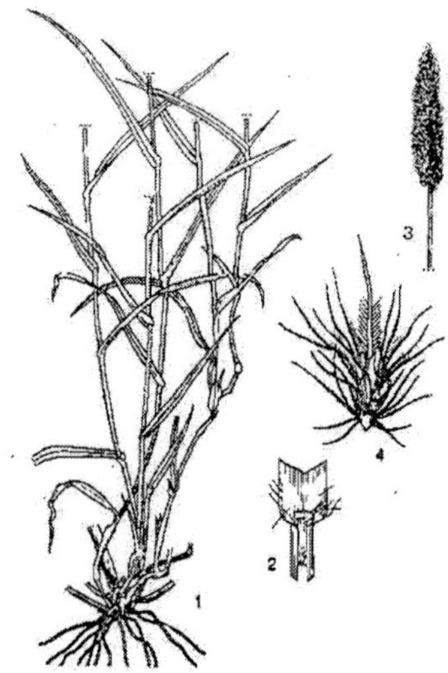

1, Habit; 2. Ligule, 3. Inflorescence, 4. Spikelet with involucre

Fig. 6.3: Habit and floral structures of Cenchrus ciliaris

BOTANICAL FEATURES OF SUGARCANE

Sugarcane: *Saccharum officinarum* $2n = 80$ **Family:** Poaceae

- It is important source of sugar.
- It is being cultivated in India in almost all the states.
- Uttar Pradesh ranks first in both area and production.

Habitat	:	Subtropical, Tropical
Habit	:	Perennial grass growing erect and tall, upto 6-8 metres, also called giant grass.
Root	:	Seminal root produced from radicle is later replaced by adventitious root which arise from basal node of the stem. Stilt roots arise from 2-3 nodes above ground level which give anchorage to the plant.
Stem	:	It is called cane or culm. Erect and unbranched, more or less cylindrical, solid with waxy coating. Internodal length is shorter at base and apex of stem and longer in the middle. Nodes are provided with root primordial and intercalary meristematic tissue.
Leaves	:	Simple, sessile, sheathed leaf base which encloses internode, alternate, distichous, ligulate, Linear leaf shape with prominent midrib and parallel venation, Leaf margin and leaf surface is quite rough.

Inflorescence	:	Terminal silky panicle called 'Arrow'. The main axis carries a number of lateral branches repeatedly branches and ultimately ending in rachii which bear the pair of spikelets with ring of long hairs below each spikelet.
Flower	:	Of the paired spikelets, one is sessile and the other one is pedicellate. Both are identical in size, shape and structure. Spikelets are deciduous. Each spikelet has two glumes surrounded by a basal ring of silk hairs which are longer than spikelets. Each spikelet has two florets inside. **Lower floret:** It is sterile represented by only lemma **Upper floret:** Bisexual, zygomorphic, Lemma is absent and Palea is thin and folded inside the lemma of sterile floret. **Lodicules** – Two **Androecium** – Three versatile anthers, Polyandrous **Gynoecium** – Monocarpellary, Superior ovary with 2 plumose stigma
Floral formula	:	As in fig.6.4
Floral diagram	:	As in fig.6.4

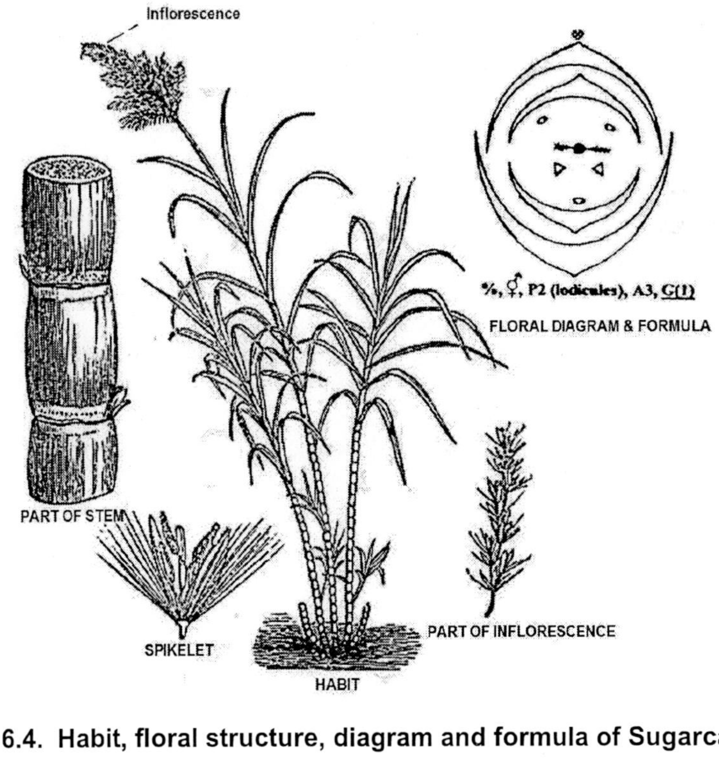

Fig. 6.4. Habit, floral structure, diagram and formula of Sugarcane

Centre of Origin: New Guinea

Putative parents: Sugarcane must have arise from *Saccharum robustum* and introgression also might have occurred from *Erianthus maximus*.

- *Saccharum spontaneum*: (2n = 40-128), A perennial grass, free tillering with aggressive rhizome. Culm is hard pithy with little juice. Has resistant to most of the major diseases.
- *Saccharum robustum*: (2n = 60-194, usually 80), It is a vigorous perennial grows along the river bank.
- *Saccharum barberi*: (2n = 81-124), North Indian indigenous, originated by natural hybridization between *S. spontaneum and S. officinarum.*
- *Saccharum sinense*: (2n = 118), Chinese cane are tall, hardy and vigorous, high fibre content and poor juicy quality
- *Saccharum officinarum*: (2n = 80), It is also called noble cane, thick cane with high sucrose
- *Saccharum edule*: (2n = 70-120), Edible inflorescence.

Related Genera of Sugarcane

Erianthus sp	(2n : 20, 30, 40 & 60) found in the canal side of South India
Miscanths flovidulus	(2n : 36 & 72)
Narenga	(2n : 30)
Sclerostachya	(2n : 30)
Imperata cylindrica	(2n : 20)

Nobilization of sugarcane

Nobilization of sugar cane was started by C. A. Baber and T. S. Venkataraman at Coimbatore. Crosses were made between unproductive canes of North India S. Barberi and tropical Noble cane *S. officinarum* (which is productive but susceptible to diseases). The hybrids then crossed to *S. spontaneum* thus transferring the vigour and disease resistnace from *S. ponstaneum* and evolved varieties suitable for North Indian conditions.

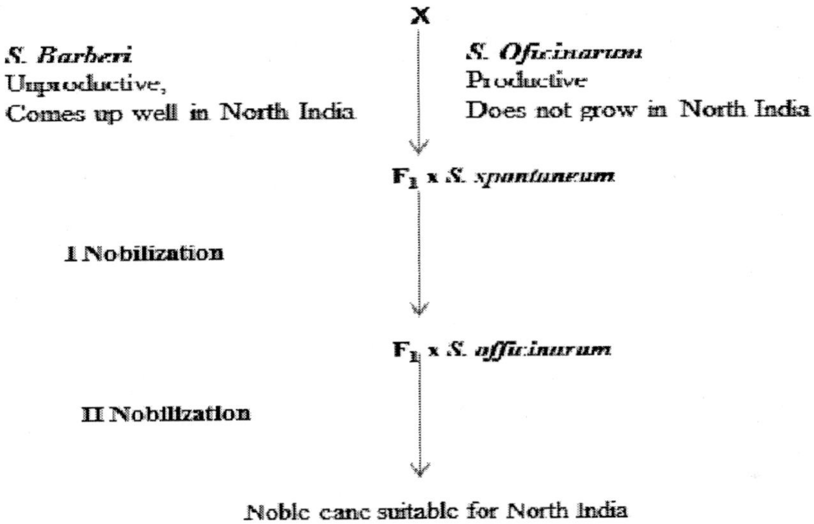

Manufacture of sugar

Sugar production involves the following steps.

1. Extraction of juice
2. Purification of the juice

1. **Extraction of juice:** Juice is extracted by crushing the cane between massive rollers. During extraction, water is sprayed on the blanket of bagasse as it emerges from each mill, to leach out the sugar (imbibitions). The final left out bagasse contains the unextracted sugar, the woody fibre and 45 to 55% water. This is used in paper manufacture or as fuel.
2. **Purification of the juice**
   (a) **Clarification:** The dark green juice from the mills is acidic and turbid. Clarification is done to remove both soluble and insoluble impurities by employing calcium oxide (lime) and heat (0.5 kg CaO per tones of cane). Lime neutralizes the natural acidity of the juice, forming insoluble lime salts, mostly calcium phosphate. Heating the lime juice to boiling coagulates the albumin and some of the fats, waxes and gums and the precipitate thus formed entraps suspended solids as well as finer particles. Mud is separated from the clear juice by sedimentation. Mud is filtered or rotary drum vacuum filters. The clear dark brown clarified juice poured to the evaporation without treatment.

   (b) **Evaporation:** Clarified juice contains about 85 per cent water. Two third of this water is evaporated in vacuum boiling bodies arranged in series so that each body has a higher vacuum, therefore boils at a lower temperature. The syrup leaving the last body contains 65 per cent solids and 35 per cent water.

(c) **Crystallization:** Crystallization is done in a single effect vacuum pans, where the syrup is evaporated until staturated with sugar. At this point seed grain is added to serve as nuclei for the sugar crystals and more syrup is added as the water evaporates. The growth of the crystals continued until the pan is full. The crystals and syrup form a dense mass known as massecuite. Then the massecuite is drawn into revolving machines called centrifuge revolving at 1000 to 1800 rpm. The output is the raw sugar.

(d) **Refining:** The raw sugar is refined by a process called affination or washing which consists of removing the adhering film of molasses from the surface of the raw sugar crystal.

(e) **Defecation:** The raw washed sugar liquor contains some insoluble materials such as bagassy, clay, sand, gums, pectins, and some true colloids. Defecation is done by

(i) **Chemical treatment:** Treatment with phosphoric acid or a soluble phosphate with lime and heat. The colour of the juice is removed by phosphate treatment.

(ii) **Filtration:** Using inert filters that permit filtration under pressure.

Uses of sugarcane

1. **White sugar (Vacuum pan sugar):** Produced from sugar industry is widely used in foods, beverages and for sweetening purposes.
2. **Khandsari (Open pan sugar):** A pale yellow sugar used in foods, beverages and for sweetening purposes.
3. **Gur or Jaggery:** The juice is squeezed out by primitive methods and evaporated over an open fire until it is almost dry. The resulting sugar contains varying amounts of sucrose, with other organic and inorganic substances, dark in colour.
4. **Molasses:** One of the most important by products from the manufacture of cane sugar, used as an industrial raw material in the production of ethyl alcohol, beverages etc.
5. **Cane syrup:** Also known as fancy molasses is prepared from sugar cane in which inversion of the juice is produced by mineral acids or invertase.
6. **Bagasse:** The fibrous residue left after the extraction of the juice from the sugar cane is the main source of fuel in sugar factories. Also used in paper making, cardboard, fibreboard, and wallboard.
7. **Furfural:** May be prepared from bagasse.
8. **Press must or filter cane:** Used as manure.
9. **Sugarcane wax:** Used in the manufacture of carbon paper, shoe and other polishes, wax paper.
10. **Cane trash and cane tops:** Used feed to cattle and elephants. Molasses is used as an animal feed.

11. The cane is partly used for chewing purpose.
12. Fresh juice is extraction and marketed.

Chemical Composition

Millable cane contains 26% organic matter (of which 9-16% fibre and 10-16% sucrose). The juice 85-88% of total w eight of the cane, is essentially water and sucrose, with some glucose, fructose and small quantities of mineral salts and nitrogenous substance. The juice brix is the total solids in the juice, which is usually 20-22% of which 80-95% is sucrose. Over all recovery of raw sugar per 100 tons or millable cane is usually 9-13 tons.

Final raw molasses contains: Water 20%, sucrose 35%; reducing sugar 14%. Bagasse contains water 45-55%, Fibre 42-47%, Sucrose 2.3-3.0 per cent.

QUESTION BANK: UNIT-II

BOTANICAL DESCRIPTION AND ECONOMIC USES OF POACEAE

PART-A

CHOOSE THE BEST 0.5 MARK

					Key
1	Cereal grains are rich in				
	a) Carbohydrate	b) Protein	c) Fat	d) Fibre	a
2	Most of the poaceae family plants having fruit types is				
	a) Drupe	b) Regma	c) Caryopsis	d) Utricle	c
3	The chromosome number for Rice is				
	a) 2n : 20	b) 2n: 24	c) 2n: 26	d) 2n: 28	b
4	The number of anthers in rice floret are				
	a) 3	b) 6	c) 4	d) 5	b
5.	The center of origin of rice is				
	a) South East Asia	b) Central Asia	c) Ethiopia	d) Asia minor	a
6.	The tropical race of Asian rice belongs to the sub species				
	a) Indica	b) Japonica	c) Javanica	d) All of then	a
7.	The alcoholic beverages manufactured from fermented rice grain is				
	a) Rice beer	b) Sake	c) Bourbon Whiskey	d) Both a & b	d
8.	The merits of parboiling of rice are				
	a) Less breakage	b) Better keeping quality	c) Preserve vitamins	d) All of them	d
9.	The inflorescence of wheat is called				
	a) Distichous spike	b) Digitate spike	c) False spike	d) Tassel	a
10.	The center of origin of hexaploid wheat is				
	a) South East Asia	b) Central Asia	c) North Africa	d) Asia minor	b
11.	The center of origin of diploid wheat is				
	a) South East Asia	b) Central Asia	c) North Africa	d) Asia minor	d
12.	Botanical name of hexaploid wheat is				
	a) Zea mays	b) Oryza sativa	c) T. aestivum	d) S. bicolor	c
13.	The protein content of wheat flour is				
	a) 10 %	b) 40 %	c) 20 %	d) 30 %	a
14.	The protein in wheat is called				
	a) Ricin	b) Zein	c) Gluten	d) Triticin	b
15.	The distilled alcoholic beverages produced from wheat grain is				
	a) Vodka	b) Wheat beer	c) Rice beer	d) Sake	a
16	The alcoholic beverages produced from fermented of wheat grain is				
	a) Vodka	b) Wheat beer	c) Rice beer	d) Sake	b

17. Botanical name of maize is
a) Zea mays b) Oryza sativa c) T. aestivum d) S. bicolor **a**
18. Male inflorescence of Maize is called
a) Cob b) Tassel c) Silk d) Ovary **b**
19. Female inflorescence of Maize is called
a) Cob b) Tassel c) Silk d) Ovary **a**
20. The protein in maize is called
a) Ricin b) Zein c) Gluten d) Eleusinin **b**
21. The distilled alcoholic beverages manufactured from maize is
a) Vodka b) Sake c) Bourbon Whiskey d) Rum **c**
22. Chromosome number of maize and sorghum is
a) 2n: 20 b) 2n: 24 c) 2n: 14 d) 2n: 25 **a**
23. Botanically sweet corn is called
a) Zea mays var. saccharata b) Zea mays var. amylacea
c) Zea mays var. indurata c) Zea mays var. ceratina **a**
24. Pod corn is chiefly used for
a) Making flour b) Making Bread c) Breeding purposes d) Textile sizing **c**
25. The number of anthers in sorghum plant are
a) 3 b) 6 c) 4 d) 2 **a**
26. Which of the following sorghum wild species are utilized for forage sorghum improvement?
a) S. halapense b) S. sudanense c) S. nitidum d) Both a & b **d**
27. The broom corn is botanically called
a) S. halapense b) S. sudanense c) S. nitidum a) S. doncha **a**
28. The origin of sorghum is
a) India b) Indonesia c) Ethiopia d) Uganda **c**
29. The chromosome number of pearl millet is
a) 2n: 14 b) 2n: 20 c) 2n: 24 d) 2n: 80 **a**
30. The condition of flowers, ovule mature the before the anther is known as
a) Protandry b) Protogyny c) Both a & b d) None of them **b**
31. The condition of flowers, anther mature the before the ovule is known as
a) Protandry b) Protogyny c) Both a & b d) None of them **a**
32. The Pearl millet is also called
a) Bulrush millet b) Cattail millet c) Spiked millet d) All of them **d**
33. The center of origin of finger millet (Ragi) is
a) India b) Indonesia c) Ethiopia d) Uganda **d**
34. Fruit of Ragi is botanically called
a) Drupe b) Utricle c) Caryopsis d) Regma **b**

35. Botanical name of foxtail millet or Italian millet is

a) S. italica b) P. purpureum c) P. milliaceum d) P. sumatranse **a**

36. Botanically sugarcane inflorescence is called as

a) Spike b) Arrow c) Spadix d) Tassel **b**

37. Botanically Nobel cane called

a) S. officinarum b) S. barberi c) S. sinense d) S. spontaneum **a**

38. Botanically Indian cane called

a) S. officinarum b) S. barberi c) S. sinense d) S. spontaneum **b**

39. Chromosome number of S. officinarum is

a) 2n: 14 b) 2n: 20 c) 2n: 24 d) 2n: 80 **d**

40. The alcoholic drinks manufactured from Sugarcane molasses is

a) Rum b) Whiskey c) Beer d) Vodka **a**

41. Nobilization of Sugarcane in India started at

a) Coimbatore b) Hyderabad c) New Delhi d) Cochin **a**

42. Which of the following crop mostly used as permanent pasture?

a) Buffel grass b) Napier grass c) Guinea grass d) none **a**

43. Which of the following is ideal for cut and carry forage grass?

a) Buffel grass b) Napier grass c) Guinea grass d) none **c**

44. The fibre content of ragi grain is

a) 10 % b) 5 % c) 3% d) 8% **c**

45. Maize is a -------------- crop

a) Monoecious b) Dioecious c) Hermaphrodite d) Self pollinated **a**

46. Inflorescence of ragi is called as

a) Arrow b) Spikelets c) Panicle d) Terminal digitate spike **d**

FILL IN THE BLANKS

1. The botanical name of rice is ------------------
2. The chromosome number of rice is --------------
3. Chromosome number of hexaploid wheat is-------------
4. Sorghum belongs to ------------- family
5. Botanical name of flour corn is ------------------
6. The center of tetraploid wheat is -------------
7. --------------g of HCN is sufficient to kill a cow
8. Poison present in sorghum plant is --------------------
9. The origin of pearl millet is ----------------

10. The botanical name of pearl millet is ----------------
11. The condition of flowers, ovule mature the before the anther is known as -------
----
12. The condition of flowers, anthers mature the before the ovule is known as ------
----
13. The botanical name of finger millet (ragi) is -----------------
14. The botanical name of little millet (samai) is --------------------
15. The botanical name of kodo millet (varagu) is ------------------
16. The botanical name of proso millet (pani varagu) is ---------------
17. Pearl millet grains are high in amino acid ------------- and --------------
18. Sugarcane inflorescence is called --------------

Answer key

1. Oryza sativa
2. $2n = 24$
3. 2n = 42
4. Poaceae
5. Zea mays var. amylacea,
6. Abyssinia
7. 0.5g
8. HCN,
9. West Tropical Africa,
10. Pennisetum glaucum
11. Protogyny
12. Protandry
13. Eleusine corocana
14. Panicum sumatranse
15. Paspalum scrobiculatum
16. Panicum milliaceum
17. Tryptophan and cysteine
18. Arrow.

MATCH THE FOLLOWING

1. Protandry	:	(a) Pearl millet
2. Protogyny	:	(b) Maize
3. Utricle	:	(c) Cumbu Napier
4. 2n: 24	:	(d) Ragi
5. Forage purposes	:	(e) Rice

Answer key

1. (b) 2. (a) 3. (d) 4. (e) 5. (c)

1. Rum	:	(a) Maize
2. Bourbon whisky	:	(b) Sugarcane molasses
3. Sake	:	(c) African beer
4. Vodka	:	(d) Rice
5. Sorghum	:	(e) Wheat

Answer key

1. (b) 2. (a) 3. (d) 4. (e) 5. (c)

1. Napier grass	:	(a) Sorghum bicolor
2. Guinea grass	:	(b) Cenchrus ciliaris
3. Buffel grass	:	(c) Echinochola frumentosa
4. Barnyard millet	:	(d) Megathyrus maximus
5. Sorghum	:	(e)Pennisetum purpureum

Answer key

1. (e) 2. (d) 3. (b) 4. (c) 5. (a)

STATE TRUE OR FALSE

1. Unisexual flower possesses both androecium and gynoecium.
2. The stem of rice also called as culm.
3. Rice is stable food for Asian countries.
4. Rice floret has three stamens.
5. The fruit type of rice is caryopsis.
6. The fruit type of sorghum is utricle.
7. Wheat floret has six stamens.
8. The botanical name of pearl millet is sorghum bicolor.
9. The alcoholic drinks rum is prepared from sugarcane molasses.
10. The fruit type of rice is Utricle.
11. Bourbon Whisky obtained from maize grains.
12. The condition of flowers, anthers mature the before the ovule is known as protandry.
13. The condition of flowers, ovule mature the before the anther is known as protandry.
14. Maize is an example for monoecious plant.
15. Sugarcane chromosome number is 80 .
16. Ragi grains are rich in Iron and Phosphorus.
17. Pearl millet seeds are rich in tryptophan and cystine.
18. The botanical name of wheat is Zea mays.
19. The chromosome number of maize is 24.
20. Finger millet is an example of famine food.

Answer keys

1. True	2. True	3. True	4. False	5. True	6. False
7. False	8. False	9. True	10. True	11. True	12. True
13. False	14. True	15. True	16. True	17. True	18. False
19. False	20. True	21. True	22. False.		

PART-A

1 MARK

1. What is mean by protandry and protogyny? Give example for each.
2. Write short notes on barboiling of rice and their merits.
3. Write short notes on sorghum poisoning.
4. What is mean by small millets? List out the small millets.
5. Uses of guinea grass.
6. Write about cumbu napier hybrid.
7. Write chemical composition of Rice.
8. Write characteristic features of Pearl millet.
9. List out the sorghum land races available in Tamil Nadu.
10. Differentiate cereals and pulses.

PART-B

2 MARK

1. Elucidate economic importance of rice.
2. Write classification of rice based on the quality traits.
3. Enumerate the origin of hexaploid wheat.
4. Write uses of wheat.
5. Write about nobilization of sugarcane.
6. Write uses of Sugarcane.
7. Classification of sorghum based on coverage of glume on the grain.
8. What are the characteristics features of poaceae family?
9. Write any two cereal crops with botanical name, chromosome number, and economic importance.
10. List out the minor millets with botanical name, chromosome number and economic part.
11. Write the uses of sorghum and pearl millet.
12. Elucidate the economic importance of maize.
13. Explain the classification maize based on kernel or endosperm structure.

PART-C

5 MARK

1. Write the economic importance of rice and maize.
2. Explain the classification of maize based on the grain structure with suitable diagrams.
3. Explain the classification of sorghum by Harlan and de wet (1970).
4. Write economical importance of Sugarcane.
5. Utilization of pearl millet and wheat ad sorghum.
6. Explain botanical description of rice.
7. What are the characteristics features of poaceae family? Write any three important crops with chromosome number and economic importance

Chapter

7

List of cultivated crops, economic parts, chromosome number, and family description of Fabaceae; Key botanical features of Red gram, Bengal Gram and Soybean

LIST OF CULTIVATED CROPS UNDER FAMILY FABACEAE

Sl. No.	List of cultivated crops	Scientific name	Chromosome number	Economic Parts
I	**Pulses**			
1	Red gram	*Cajanus cajan*	22	Seed
2	Bengal gram	*Cicer arietinum*	16	Seed
3	Soybean	*Glycine max*	40	Seed
4	Blackgram	*Vigna mungo*	22	Seed
5	Green gram	*Vigna radiata*	22	Seed
6	Cowpea	*Vigna unguiculata*	22	Seed
7	Lab lab	*Lablab purpureus*	22	Seed
8	Horse gram	*Macrotyloma uniflorum*	24	Seed
II	**Oil seeds**			
9	Ground nut	*Arachis hypogaea*	40	Seed
III	**Forage legumes**			
10	Lucern / Alfalfa	*Medicago sativa*	16,32,64	Leaves & toppings
11	Stylosanthus	*Stylosanthus hamata*	40	Leaves & toppings
12	Clitoria	*Clitoria ternatea*	16	Leaves & toppings
IV	**Tree fodder**			
13	Agathi	*Sesbania grandiflora*	24	Leaves & toppings
V	**Green manure**			
14	Sunnhemp	*Crotalaria juncea*	16	Whole plant
15	Sesbania / Daincha	*Sesbania aculeata*	12,24	Whole plant

Family description of Fabaceae

Plant	:	Annual / perennial, Herb / Shrub / Tree
Stem	:	Branched, herbaceous / woody
Leaf	:	Compound leaf (Except sunnhemp which has simple leaf), stipulate
Inflorescence	:	Terminal / axillary raceme
Flower	:	Pedicellate, bracteate, zygomorphic, bisexual, complete, papilionaceous, pentamerous, hypogynous
Calyx	:	Sepals - five, gamosepalous, valvate aestivation
Corolla	:	Petals - five, polypetalous, vexillary aestivation
Androecium	:	Stamens – Ten, Diadelphous (Except Soybean, Ground nut & Sunnhemp which have Monadelphous androecium)
Gynoecium	:	Monocarpellary, superior ovary, unilocular with marginal placentation
Fruit	:	Legume / Pod, seed is non-endospermic

Systematic position of family Fabaceae

Kingdom	:	Plant kingdom
Division	:	Phanerogams
Subdivision	:	Angiosperms
Class	:	Dicotyledonae
Subclass	:	Polypetalae
Series	:	Calyciflorae
Order	:	Rosales
Family	:	Fabaceae

Characteristics of Pulses

- Pulses contain 17 – 40 % protein.
- They contain 1 – 2 % fat.
- They are rich in Lysine, Threonine & Tryptophan but low in sulphur containing aminoacids viz., Methionine, Cysteine & Cystine.
- They have root nodules which fix up atmospheric nitrogen into the soil.
- Many grain legumes contain toxic substance like trypsin inhibitors which are removed by cooking.

BOTANICAL FEATURES OF RED GRAM

Redgram (Pigeonpea, Arhar) – *Cajanus cajan* $2n = 22$ **Family:** Fabaceae

Habitat	:	Tropical
Habit	:	Annual or perennial shrub erect and branching
Root	:	Tap root systems. Large clusters of nodules are produced under favourable condition
Stem	:	Branching, woody, cylindrical and hairy
Leaves	:	Trifoliately compound, central leaflet longer than lateral ones, Petiole grooved and long, stipulate, pulvinus, Leaf lamina hairy, lanceolate leaf shape, Entire leaf margin, reticulate venation
Inflorescence	:	Axially or terminal racemes on long peduncle.
Flower	:	Bracteates, zygomorphic, bisexual, complete, papilionaceous, pentamerous
Calyx	:	Sepals – Four lobed and two being united
Corolla	:	Petals – five (One standard, Two wing & two keel petals), Polypetalous, Papilionaceous, Vexillary aestivation
Androecium	:	Stamens – Ten, Diadelphous
Gynoecium	:	Monocarpellary, superior ovary, many ovules in marginal placentation, capitate stigma
Fruit	:	A pod. Flattened with constrictions between seeds
Pollination	:	Often cross pollination
Floral Formula	:	As in fig.7.1
Floral Diagram	:	As in fig.7.1

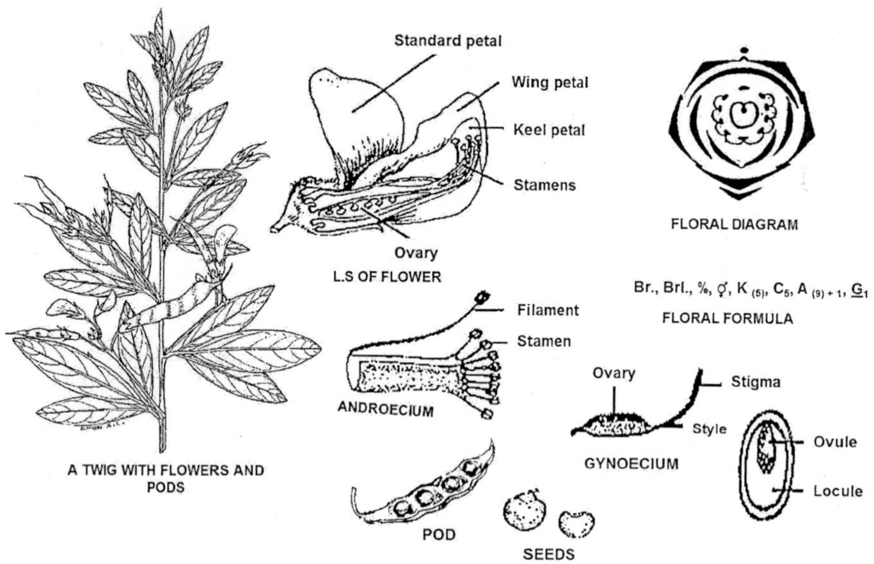

Fig. 7.1: Morphology, floral structure and diagram of Cajanus cajan

Origin: Africa

Cultivated forms of Redgram

1. *Cajanus cajan* var. *flavus*
2. *Cajanus cajan* var. *bicolour*

Cajanus cajan var. flavus	*Cajanus cajan var. bicolour*
• Early maturity & Short plant	• Late maturity & Bushy plant
• Standard petal – Yellow colour	• Standard petal – Red / Purple colour
• Green pods	• Blotched with Maroon
• Usually 3 seeded	• Usually 4 – 5 seeded
• Tur cultivars of India	• Arhar cultivars of India

Economic Importance

- The green pods and green seeds are eaten as vegetable. The ripe dry seeds are boiled and eaten as pulse.
- Dried husk and broken pieces of seed are used as cattle feed.
- Green leaf and tops of the plant is used as fodder and for making silage.
- Dried stake is used as fuel and as thatching material.

BOTANICAL FEATURES OF BENGAL GRAM

Bengalgram (Chickpea, Channa) - *Cicer arietinum* $2n = 16$ **Family:** Fabaceae

Habitat	:	Subtropical and temperate
Habit	:	Herbaceous annual, branching from the base, mostly erect and with a few spreading branches, 25-20 cm tall, all parts of the plant are covered with glandular hairs.
Root	:	Tap root with nodules
Stem	:	Greyish in appearance. Main stem is rounded, branches are usually quadrangular ribbed and green
Leaves	:	Imparipinnate, leaflets hairy, serrated leaf margin, ovate shape, stipulate
Inflorescence	:	Solitary, axillary, pedicels jointed.
Flower	:	Bracteates, zygomorphic, bisexual, complete, papilionaceous, pentamerous
Calyx	:	Sepals - five, gamosepalous, valvate aestivation
Corolla	:	Petals – five, polypetalous, papilionaceous, vexillary aestivation
Androecium	:	Stamens – Ten, diadelphous
Gynoecium	:	Monocarpellary, superior ovary, capitate stigma
Fruit	:	Pod, one or two seeded. Seeds angular with prominent beak and small hilum. Colour ranging from white red to black. Cotyledons are thick and yellowish
Pollination	:	Self pollination
Floral Formula	:	As in fig.7.2
Floral Diagram	:	As in fig.7.2

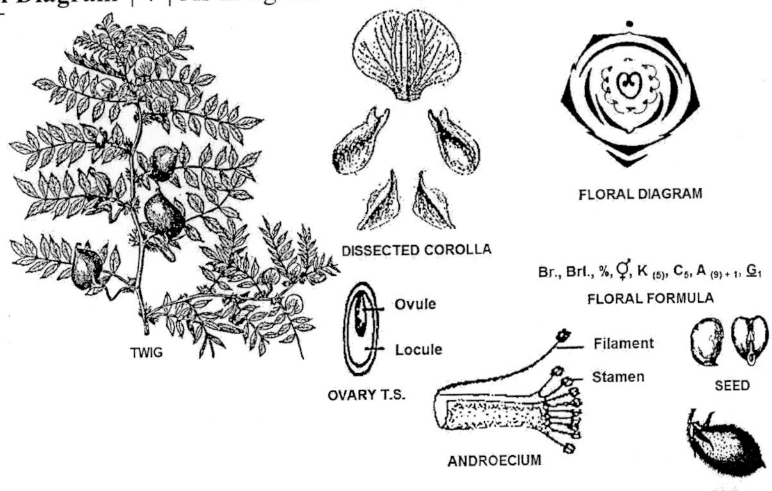

Fig. 7.2: Bengal gram (Cicer arietinum) floral structure, diagram, and formula

Origin: Tropical Asia

Putative parent: Not known

Related species: C. chorossanicum, C. cuneatum, C. pinnatifidum

Cultivated types of Bengal gram

Two broad groups viz., 1. Desi, 2. Kabuli

Desi	Kabuli
• Plants are dwarf with dark green leaves • Flowers are purple in colour • Seeds are dark brown coloured, wrinkled & smaller in size	• Plants are tall with pale green leaves • Flowers are white in colour • Seeds are white coloured, smooth & larger in size

Economic Importance

- Dried seeds soaked in water, cooked and eaten.
- Dhal used in various food preparations.
- Popped grain – ‘ Pottu kadalai’
- Broken seed as dhall – ‘ Kadalai paruppu’
- Flour used in various food preparations.
- Roasted seeds are taken as food and also used in preparation of various dishes.
- Green pods and tender shoots used as vegetable. Dried plant as cattle feed.
- An acid liquid from glandular hairs is collected by spreading a thin cloth over night. The collected liquid contains 94 per cent malic acid and 6 per cent oxalic acid. It is used as vinegar and also as medicine.

BOTANICAL FEATURES OF SOYBEAN

Soybean : *Glycine max* ($2n = 40$) **Family:** Fabaceae

Habitat	:	Subtropical and tropical
Habit	:	Herbaceous annual, pubescent (Dense hairs on all parts of body)
Root	:	Tap root with nodules
Stem	:	Sub erect or climbing
Leaves	:	Trifoliate, long petioled, leaflects ovate shape, acute apex
Inflorescence	:	terminal raceme in determinate plant Axillary raceme in indeterminate plant
Flower	:	Small, Bracteates, Zygomorphic, Bisexual, Complete, Papilionaceous, Pentamerous
Calyx	:	Sepals - five, Gamosepalous, Valvate aestivation

Corolla	:	Petals – five, Polypetalous, Papilionaceous, Vexillary aestivation
Androecium	:	Stamens – Ten, Monadelphous, uniform anthers, globose
Gynoecium	:	Monocarpellary, Superior ovary, Capitate stigma
Fruit	:	Pods linear to oblong profusely hairy. The seed of soybean is oval with the colour varying from light yellow to black. The hilum is prominent. Seed viability is poor
Pollination	:	Self pollination
Floral Formula	:	As in fig.7.3
Floral Diagram	:	As in fig.7.3

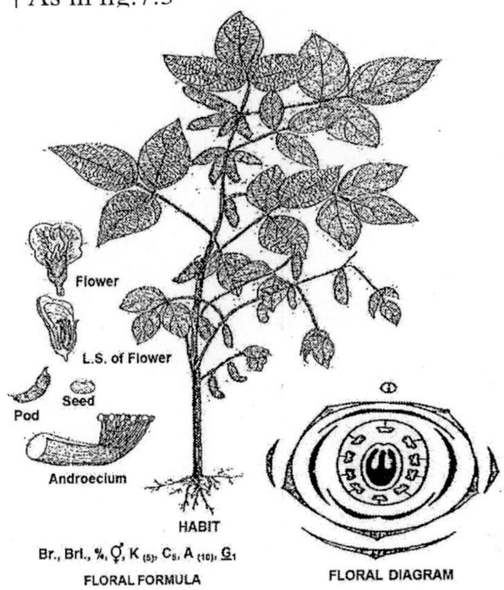

Fig. 7.3: Soya bean (*Glycine max*), habit, floral structure and diagram

Origin: China

Wild relatives: G. max derived from G. ussutiensis and hybridization with G. tomentosa

Cultivated types of Soybean

1. Yellow seeded form - Oil purpose
2. Green seeded form - Vegetable purpose
3. Brown seeded form - Fodder purpose
4. Black seeded form - Rich in protein

Economic Importance

- Most important source of oil and protein.
- Unripe seeds are used as vegetable. Dried seed are eaten whole.

- Substitute for black gram in preparation of flour for idli.
- Soya milk, soya sauce, soya oil are used for edible purpose.
- The whole plant as fodder and seed as cattle feed.
- Soya meal or protein used to manufacture synthetic fibre, adhesive, fire fighting foam.
- Soya flour used in bakery.

Chapter

8

Key botanical features of Black gram, Green gram, Cowpea, Lablab, Horse gram and Ground nut

BOTANICAL FEATURES OF BLACKGRAM

Blackgram (Urdbean) *Vigna mungo* $2n = 22$ **Family:** Fabaceae

Habitat	:	Tropical
Habit	:	Annual, herbaceous, 30-100 cm in height, erect, semi-erect to trailing or spreading types, plant densely hairy
Root	:	Tap root with nodules
Stem	:	Slightly woody, covered with brown hairs, branched from the base
Leaves	:	Trifoliate, alternate, petiolate, stipulate, pulvinate, leaflets ovate / Lanceolate, entire, acute, sparsely hairy on both surfaces, reticulate venation
Inflorescence	:	Axillary raceme
Flower	:	Bracteate, pedicellate, zygomorphic, bisexual, hypogynous, complete, pentamerous
Calyx	:	Five sepals, gamosepalous, valvate aestivation
Corolla	:	Five petals, polypetalous, vexillary aestivation
Androecium	:	Ten stamens, Diadelphous (9 + 1)
Gynoecium	:	Monocarpellary, superior ovary with marginal placentation
Fruit	:	Pod Seed: Oblong with square ends, black. Hilum white and concave, seed coat surface is smooth. Cotyledons are white in colour
Pollination	:	Self pollination
Floral Formula	:	As in fig.8.1
Floral Diagram	:	As in fig.8.1

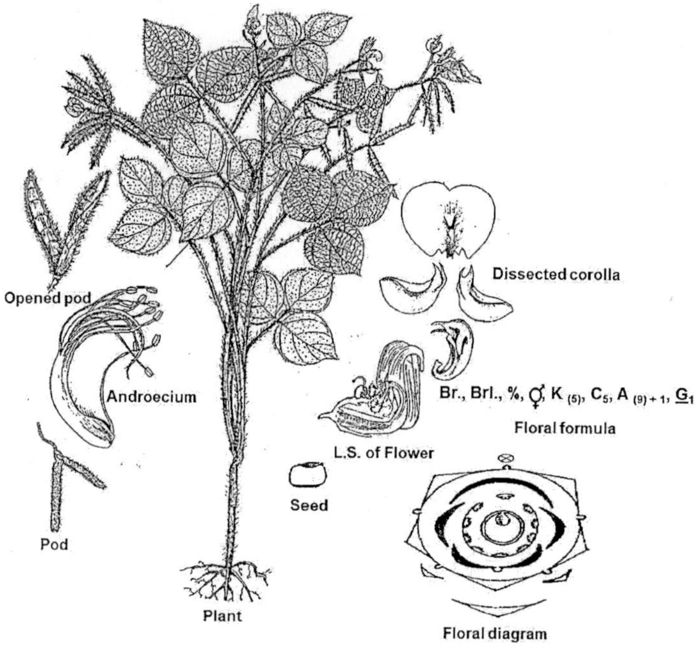

Fig. 8.1: Black gram (Vigna mungo) habit, floral structure, diagram and formula

Origin: India

Putative parent: *Vigna sublobata*

Economic Importance

- Dried grain used in the preparation of idli, dosai, vadai etc.
- Flour is used in bakery for preparation of bread and biscuit.
- Green seeds and pods are eaten as vegetable.
- Dried or green plants are used as fodder.
- Broken grain, seed coat and bhusa (pottu) are excellent fodder.

Difference between Blackgram and Greengram

Sl.No.	Characters	Blackgram	Greengram
1	Plants	Densely hairy plants	Sparsely hairy plants
2	Leave colour	Pale green colour	Dark green colour
3	Petal colour	Pale yellow colour	Yellow colour
4	Pod	Short	Long
5	Seed	Oblong with square ends	Globular
6	Testa	Black in colour	Green in colour
7	Hilum	White and concave	White and round
8	Cotyledon	White in colour	Yellow in colour

BOTANICAL FEATURES OF GREENGRAM

Green Gram (Mungbean) *Vigna radiata* (2*n* = 22) **Family:** Fabaceae

Habitat	:	Tropical
Habit	:	Herbaceous annual with slight tendency for twining in the upper branches, plant sparsely hairy.
Root	:	Tap root system with nodules
Stem	:	Erect or semi erect and green in colour
Leaves	:	Trifoliate with long petioles, Stipulate, leaflets entire, ovate, lobed leaves are also present
Inflorescence	:	Axillary racemes
Flower	:	Bracteate, pedicellate, zygomorphic, bisexual, hypogynous, complete, pentamerous
Calyx	:	Five sepals, gamosepalous, valvate aestivation
Corolla	:	Five petals, polypetalous, vexillary aestivation
Androecium	:	Ten stamens, Diadelphous (9 + 1)
Gynoecium	:	Monocarpellary, superior ovary with marginal placentation
Fruit	:	Pod Seed: Globular, green and the surface has fine wavy ridges. Hilum is white and flat cotyledon is yellowish in colour
Pollination	:	Self pollination
Floral Formula &	:	As in fig.8.2
Floral Diagram	:	

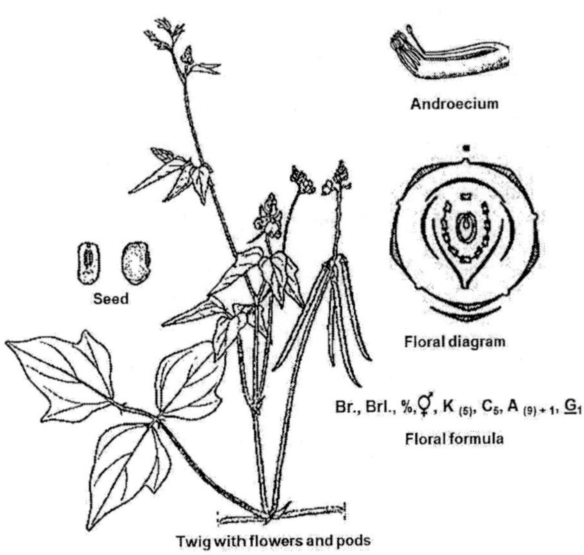

Fig. 8.2. Green gram: Vigna radiata (Papilionaceae)

Economic Importance

Seed as food for human beings, green and dry plant as fodder.

Origin: India

Putative parent: Vigna sublobata

BOTANICAL FEATURES OF COWPEA

Cowpea: *Vigna unguiculata* (2*n* = 22) **Family:** Fabaceae

Habitat	:	Subtropical and Tropical
Habit	:	Annual, twining, rarely sub erect and erect herb
Root	:	Tap root with nodules
Stem	:	Branched and slightly ridged
Leaves	:	Trifoliate, alternate, stipulate, petiolate, pulvinate, leaflets ovate, entire, acute, reticulate venation
Inflorescence	:	Axillary raceme
Flower	:	Pedicellate, bracteates, zygomorphic, bisexual, hypogynous, complete, pentamerous
Calyx	:	Five sepals, gamosepalous, valvate aestivation
Corolla	:	Five petals, polypetalous, vexillary aestivation
Androecium	:	Ten stamens, Diadelphous (9 + 1)
Gynoecium	:	Monocarpellary, Superior ovary
Fruit	:	Pod Seed: Sub reniform to subglobose
Pollination	:	Self pollination
Floral Formula	:	As in fig.8.3
Floral Diagram	:	As in fig.8.3

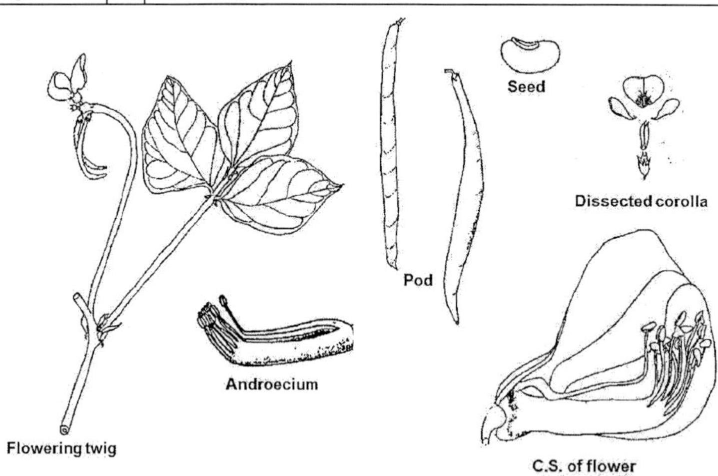

Fig. 8.3. Cowpea: Vigna unguiculata (Papilionaceae)

Origin: Tropical Africa

Related species: *Vigna unguiculata sub sp. unguiculata*
Vigna unguiculata sub sp. cylindrica
Vigna unguiculata sub sp. sesquipedalis

Economic Importance

- Tender pods are used as vegetable cowpea, tender leaves are used as greens.
- Sprouted seed as vegetable, grain as pulse.
- Whole plant as green fodder.

BOTANICAL FEATURES OF LABLAB

Lablab – *Lablab purpureus* $(2n = 22, 24)$ **Family:** Fabaceae

Habitat	:	Tropical
Habit	:	Trailing annual herb
Root	:	Tap root with nodules
Stem	:	Solid, cylindrical, herbaceous, strong twiner
Leaves	:	Trifoliate, alternate, stipulate, petiolate, pulvinate, leaflets ovate, entire, acute, reticulate venation
Inflorescence	:	Axillary raceme
Flower	:	Pedicellate, bracteates, zygomorphic, bisexual, hypogynous, complete, pentamerous
Calyx	:	Five sepals, gamosepalous, valvate aestivation
Corolla	:	Five petals, polypetalous, vexillary aestivation
Androecium	:	Ten stamens, Diadelphous (9 + 1)
Gynoecium	:	Monocarpellary, Superior ovary
Fruit	:	Pod; Seed: Non-endospermous, large size, thick, ovoid
Pollination	:	Self pollination
Floral Formula &	:	As in fig.8.4

Floral Diagram

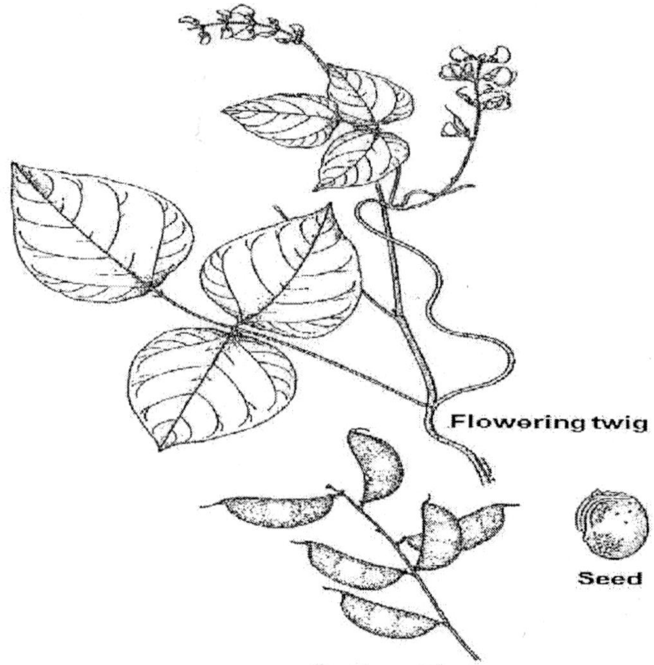

Fig 8.4: Lab lab (Dolichos lablab)

Difference between Garden Bean and Field Bean

Sl.No.	Character	Garden bean / Avarai Lablab purpureus var. typicus	Field bean / Mochai Lablab purpureus var. lignosus
1	Habit	Perennial, cultivated as on annual twiner	Annual, bushy
2	Plant parts	No pungent smell	Has pungent smell
3	Flower	White or purple	Usually white
4	Pericarp	Slender or soft	Tough & firm.
5	Seeds	Long axis of seeds are parallel to the length of the fruit	Long axis of seeds are at right angles to the length of fruit.
6	Edible part	Whole fruit	Generally seed alone.

Economic Importance: Seed as source of protein and pod as vegetable

BOTANICAL FEATURES OF HORSE GRAM

Horse Gram (Kollu) – *Macrotyloma uniflorum* $(2n = 24)$

Family: Fabaceae

Horse gram is the poor man's pulse crop in Southern India. It is a hardy, drought resistant annual grain legume.

Habitat	:	Sub – tropical, tropical
Habit	:	Slender, sub erect, profusely branched at base intertwining themselves
Root	:	Tap root with nodules
Stem	:	Slightly twining
Leaves	:	Trifoliate, alternate, stipulate, petiolate, pulvinate, leaflets ovate, entire, acute, reticulate venation
Inflorescence	:	Axillary raceme
Flower	:	Pedicellate, bracteates, zygomorphic, bisexual, hypogynous, complete, pentamerous
Calyx	:	Five sepals, gamosepalous, valvate aestivation
Corolla	:	Five petals, polypetalous, vexillary aestivation
Androecium	:	Ten stamens, Diadelphous (9 + 1)
Gynoecium	:	Monocarpellary, superior ovary
Fruit	:	Pods
Pollination	:	Self pollination
Floral Formula &	:	As in fig.8.5
Floral Diagram		

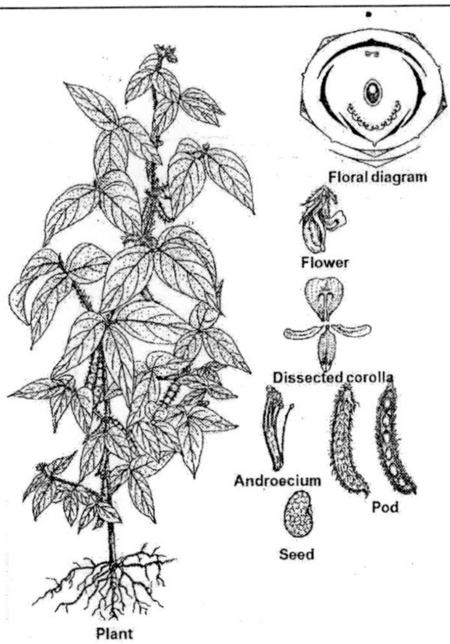

Fig.8.5. Horse gram (Macrotyloma uniflorum) habit, floral diagram

Place of Origin : South East Asia.

Distinguishing features

(i) Pubescent plant body.
(ii) Sickle shaped pod.
(iii) Compressed seed. .

Uses:

- Seeds are used as pulse. Sprouted seed used for making dishes.
- Plant and seeds are used as cattle feed
- Seeds have medicinal value, Diuretic
- Source of protein (24 per cent) used as a concentrate for cattle, horses, human food, medicinal and plant as fodder and green manure.

9. Botanical features of Groundnut

Groundnut – *Arachis hypogaea* ($2n = 40$) **Family:** Fabaceae

Habitat	:	Tropical and Subtropical
Habit	:	Low growing annual herb
Root	:	Tap root with nodules
Stem	:	Therearethreecultivatedforms(1)bunch,(2)semi-spreadingand (3) spreading types. The central axis of the stem erect, often hairy with short internodes. Some what angular slightly pithy. In bunchy types, the lateral branches also grow almost erect whereas, in the spreading (prostrate) types the laterals are procumbent or prostrate
Leaves	:	Alternate, Peripinnately compound leaves, stipulate, the petioles fairly long with two pairs of ovate leaflets, the leaflets are arranged opposite to each other in pairs, pulvinus
Inflorescence	:	Single or three to four per leaf axils in condensed raceme. The flower is sessile, yellow in colour with fairly long calyx tube which gives the false appearance of a pedicel.
Flower	:	Sessile, bracteates, zygomorphic, bisexual, hypogynous, complete, pentamerous
Calyx	:	Tubular ending in five lobes, of which three are united into one big structure and two are united separately. The long slender calyx tube is also called as Hypanthium
Corolla	:	Five petals, polypetalous, vexillary aestivation
Androecium	:	Stamens -10, monadelphous, eight fertile and two staminodes. The eight anthers are dimorphic with four having long anther lobes and four with rounded anther lobes

Gynoecium	:	Monocarpellary, superior ovary with marginal placentation
Fruit	:	An indehiscent pod carried on a long stalk which is the gynophores or carpophores. One to three seeds per pod. The testa is commonly rose or varying from dark purple to red colour or variegated
Pollination	:	Self pollination
Floral Formula	:	As in fig.8.6
Floral Diagram	:	As in fig.8.6

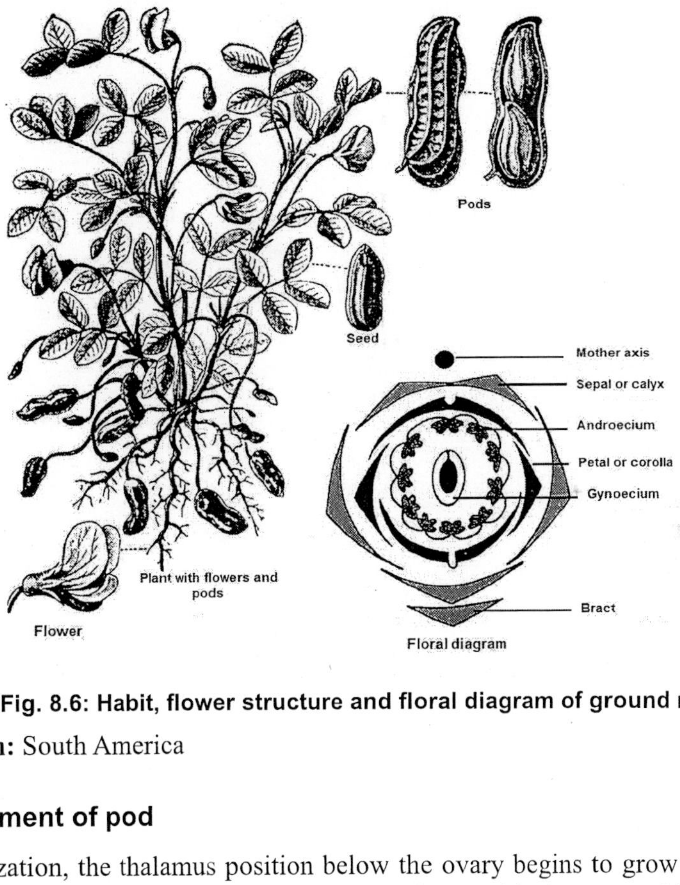

Fig. 8.6: Habit, flower structure and floral diagram of ground nut.

Origin: South America

Development of pod

On fertilization, the thalamus position below the ovary begins to grow into a peg or gynophores. The peg grows towards the soil ie., positively geotrophic, a lignified tissue forms a protective cap at the tip for the fertilized ovary. The peg carrying the ovary pushes itself into the soil. After entering the soil to a specific depth which is characteristic for a variety, it takes a horizontal position and the ovary begins to develop into a pod.

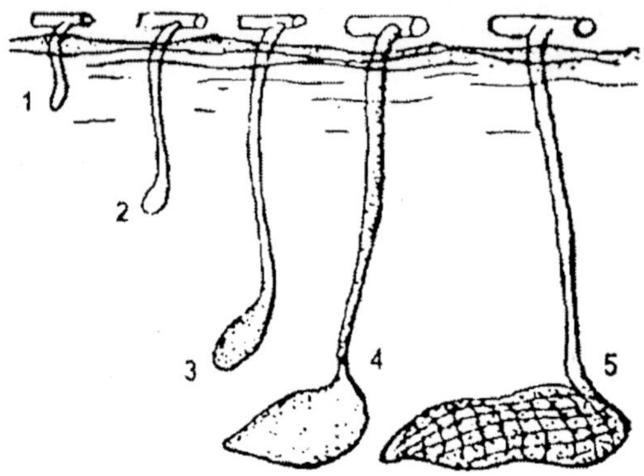

Fig. 8.7: Stages of Pod development in Ground nut

Varietal Forms

1. Arachis *hypogaea* sub sp. *hypogaea* var. *hypogaea* – Virginia runner
2. Arachis *hypogaea* sub sp. *hypogaea* var. *hirsuta* – Peruvian
3. Arachis *hypogaea* sub sp. fastigiata var. *vulgaris* – Spanish bunch
4. Arachis *hypogaea* sub sp. *fastigiata* var. *fastigiata* – Valencia

Subspecies	Variety	Type	Description
hypogaea	*hypogaea*	Virginia	No floral axis on main stem alternating pairs of floral and vegetative axes on branches, branches short less hairy.
	hirsuta	Peruvian	No floral axes in main stem alternating pairs of floral and vegetative axes on branches long more hairy
fastigiata	*vulgaris*	Spanish	Floral axes on main stem sequencial floral axes on braches more branches, upright branches.
	fastigiata	Valencia	Floral axes on main stem, sequential floral axes on branches, little branched, curved branches.

Economic Importance

- The nuts are eaten raw or after roasting. For this purpose bunch types with extra large kernals are preferred since they contain relatively low oil content than Virginia type.
- Kernel is rich source of phosphorus vitamin, protein (26%) and oil (45 to 50%).
- Oil is used as a cooking media.
- Hydrogenated oil – for preparation of vanaspathi / vegetable ghee.

- Manufacture of margarine – butter like substance.
- Oil cake is used as cattle feed.
- Moist oil cake for production of aflatoxin from Aspergilus flavus.
- Used in Pharmaceutical industry, Soap industry & Lubrication.
- New textile fibre Ardil – manufactured from peanut protein.

QUESTION BANK- UNIT: III

BOTANICAL DESCRIPTION AND ECONOMIC USES OF PAPILIONACEAE

PART-A

CHOOSE THE CORRECT ANSWER 0.5 MARK

					Key
1.	Pulses seeds are rich in				
	a) Oil	b) Carbohydrate	c) Fat	d) Protein	d
2.	Biodiesel yielding plant				
	a) Groundnut	b) Jatropha	c) Castor	d) Gingelly	b
3.	Diadelphous stamens are mostly present in family of				
	a) Poaceae	b) Malvaceae	c) Curcubitaceae	d) Fabaceae	d
4.	Which of the following having monoadelphous stamens				
	a) Soybean	b) Black gram	c) Red gram	d) Lab lab	a
5.	The leaf of Bengal gram is				
	a) Imparipinnate	b) Paripinnate	c) Trifoliate	d) None of them	a
6.	The protein and oil content of soybean is				
	a) 10 & 20 %	b) 20 & 30 %	c) 20 & 40 %	d) 20 & 50 %	c
7.	-------------------- belongs to the family papilionaceae				
	a) Red gram	b) Rice	c) Sesamum	d) Jute	a
8.	The synthetic textile fibre obtained from groundnut is				
	a) Rilson	b) Ardil	c) Fibril	d) Cellulose	b
9.	Botanical name of ground nut is				
	a) V. radiata	b) V. mungo	c) G. max	d) A. hypogea	d
10.	Chromosome number of ground nut and soybean is				
	a) 20	b) 30	c) 40	d) Both a and b	c
11.	Infloresence of most of the pulse crops is				
	a) Cyme	b) Raceme	c) Axilary raceme	d) None of them	c
12.	Protein content of bengal gram (chickpea) is				
	a) 20	b) 10	c) 17	d) 24	c
13.	Which of the following is green manure aswell as fibre crop				
	a) Cotton	b) Sunhemp	c) Agathi	d) clitoria	b

FILL IN THE BLANKS

1. Botanical name of soyabean is --------------------
2. Ground nut belongs to the family -------------------
3. --------------- crop is used as important food or feed for horses
4. Botanical name of black gram is ------------------------
5. Botanical name of red gram is -----------------------
6. Botanical name of bengal gram is ----------------------
7. Botanical name of green gram is --------------------
8. Botanical name of cowpea is ------------------------
9. Botanical name of lucern is is ---------------------
10. The common fruit type of pulses is ----------------
11. Root nodules are present in the member of --------------- family.

Answer key

1. Gycine max
2. Papilionaceae
3. Horsegram
4. Vigna mungo
5. Cajanus cajan
6. Cicer arietinum
7. Vigna radiata
8. Vigna ungiculata
9. Medicago sativa
10. Pod
11. Papilionaceae.

MATCH THE FOLLOWING

1.	Soybean	(a) Diadelphous
2.	Black gram	(b) Monoadelphous
3.	Bengalgram	(c) Green manure
4.	Sunnhemp	(d) Forage legume
5.	Stylosanthus	(e) Imparipinnate leaves

Answer key:

1. (b) 2. (a) 3. (e) 4. (c) 5. (d)

STATE TRUE OR FALSE

1. Sunhemp is example for green manure and stem fibre.
2. The economic part of sunn hemp is whole plant.
3. Most of the pulse crops have trifoliate leaves types.
4. Diadelphous androecium is occurred in ground nut.
5. Botanical name of avarai or garden bean is Dolichos lab lab var. typicus.

6. The chromosome number of redgram is 22.
7. Blackgram belongs to papilionaceous family
8. The chromosome number of ground nut is 40.
9. Botanical name of Agathi is Clitoria ternatea.

Answer key:

1. True	2. True	3. Trifoliate	4. False	5. False
6. True	7. True	8. True	9. False	

PART-B

1 MARK

1. Differentiate cereals and pulses.
2. Classification of ground nut based on the habit.
3. List out the legume forages with botanical names.
4. Differentiate garden bean and field bean.
5. Write uses of lab lab.
6. Describe a papilionaceous flower.

PART-C

2 MARK

1. Differentiate desi and kabuli types of Bengal gram.
2. Differentiate black gram and green gram.
3. Write short notes on pod or peg development in groundnut.
4. Write utilization of soybean.
5. Write the classification of cowpea.
6. Differentiate the family poaceae and papilionaceae.
7. How many stamens are present in the redgram and soybean? How does the arrangement of stamens differ between these two crops?
8. Write uses of red gram.

PART-D

5 MARK

1. What are pulses? Describe in detail botanical description of redgram.
2. List out the legume forages with botanical name. Elucidate any one of them in detail.
3. Explain distinguishing characters of papilionaceae family.
4. Describe the botanical features of ground nut.
5. Describe the botanical features of soybean.

TAMIL NADU AGRICULTURAL UNIVERSITY

Agricultural College and Research Institute, Eachangkottai
I B.Sc. (Agriculture) – I Semester (2017 Syllabus)
PBG 101. Introduction to Agricultural Botany (1 + 1)
Mid Semester Examination, 2017

Date: 30.10.2017 **Total Marks: 20**
Time: One hour

PART - A

I. Fill up the blanks (Answer all the questions) **(10 x 0.5 = 5)**

1. The basic unit of classification is ....
2. Binomial nomenclature was strictly followed by .... in his book 'Species Plantarum'
3. Fibrous root system is characteristics of ....
4. .... have stimulating effects upon drinking
5. Scientific name of rice is ....
6. Junction of leaf sheath and leaf lamina is called ....
7. Floral members are inserted above ovary is called ....
8. Rice has .... number of stamens
9. Stamens fused with petals is referred as ....
10. Lablab belongs to .... family

II. Match the following (Answer all the questions) **(10 x 0.5 = 5)**

1.	Genera Plantarum	:	Apocarpous
2.	Corolla of Fabaceae	:	Tassel
3.	Sorghum poison	:	Bentham & Hooker
4.	Fused carpels	:	Plumule
5.	Lodicules absent	:	Raceme
6.	Staminate florets	:	Syncarpous
7.	Inflorescence of sugarcane	:	Corymb
8.	Embryonic leaf	:	Cotton
9.	Fibre crop	:	Pennisetum glaucum
10.	Acropetal succession	:	Vexillary
		:	Arrow
		:	HCN

PART - B

Write short notes on any five of the following **(5 x 2 = 10)**

1. Describe the following terms with an example:
   (a) Green manure crops
   (b) Distichous leaf arrangement
   (c) Syngenecious anthers
   (d) Caryopsis fruit
2. Write the rules of ICBN.
3. Write scientific name and 2n chromosome number of the following crops:
   (a) Maize
   (b) Ragi
   (c) Redgram
   (d) Groundnut
4. Write the difference between racemose and cymose inflorescence.
5. Explain about Nobilization of sugarcane.
6. Write key botanical features of greengram.

Chapter

9

Key botanical features of Lucerne, Stylosanthes, Clitoria, Agathi, Sunnhemp and Sesbania

BOTANICAL FEATURES OF LUCERNE

It is considered as queen of fodder contains five times higher protein than fodder sorghum and high vitamin A. Cultivated species of lucerne is tertraploid

Lucerne (Alfalfa) *Medicago sativa* $2n = 16, 32,$ and 64

Family: Fabaceae

Habitat	:	Subtropical, tropical and drought tolerant
Habit	:	Deep rooted perennial, herb
Root	:	Deep penetrating tap root
Stem	:	Procumbent, arising from a woody base
Leaves	:	Trifoliate, Alternate, petiolate, Stipulate, Pubscent on lower surface, glabrous on upper surface, Leaflets are ovate, dentate and sharply toothed
Inflorescence	:	Axillary raceme with 10-35 purple flowers
Flower	:	Bracteate, pedicellate, zygomorphic, bisexual, hypogynous, complete, pentamerous
Calyx	:	Five lobed, gamosepalous, tubular with pointed teeth
Corolla	:	Five petals, polypetalous, vexillary aestivation
Androecium	:	Ten stamens, Diadelphous
Gynoecium	:	Monocarpellary with many ovules, Superior ovary with marginal placentation
Fruit	:	Spirally coiled legume
Pollination	:	Cross pollination

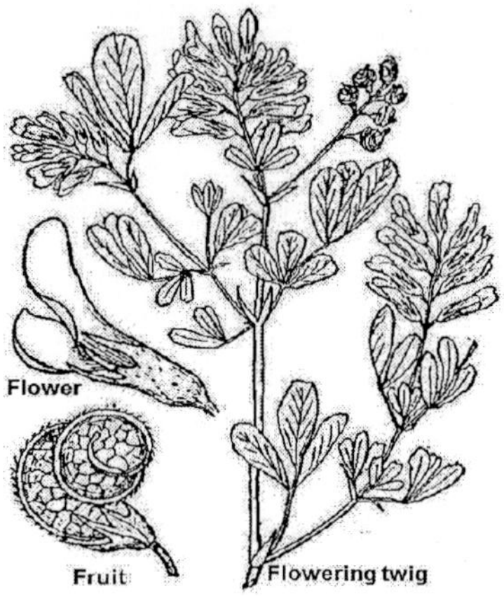

Fig. 9.1: Lucerne: (Medico sativa) flowering twig, flower and fruit

Origin: Mediterranean region

Related species: Medicago falcata

Economic Importance

- It is a very good fodder crop with 22-25% protein with well balanced calcium and phosphorus.
- Dried leaves are powdered and used in the preparation of concentrates.
- It is also used in poultry feed.
- It helps in soil conservation as it has penetrating roots.

BOTANICAL FEATURES OF STYLOSANTHUS

Stylosanthus (Muyal masal) *Stylosanthus hamata* ($2n = 22, 44, 66$)
Family: Fabaceae

Habitat	:	Tropical
Habit	:	Shrub, with many branches
Root	:	Tap root system
Stem	:	Semi erect, branching dichotomously
Leaves	:	Trifoliate, Stipulate, adnate to the base of the petiole
Inflorescence	:	Axillary / terminal racemes
Flower	:	Bracteate, pedicellate, zygomorphic, bisexual, hypogynous, complete, pentamerous
Calyx	:	Five sepals, gamosepalous
Corolla	:	Five petals, polypetalous, vexillary aestivation
Androecium	:	Ten stamens, Monadelphous

Gynoecium	:	Monocarpellary, Superior ovary with marginal placentation
Fruit	:	Two segmented lomentum
Pollination	:	Self pollination

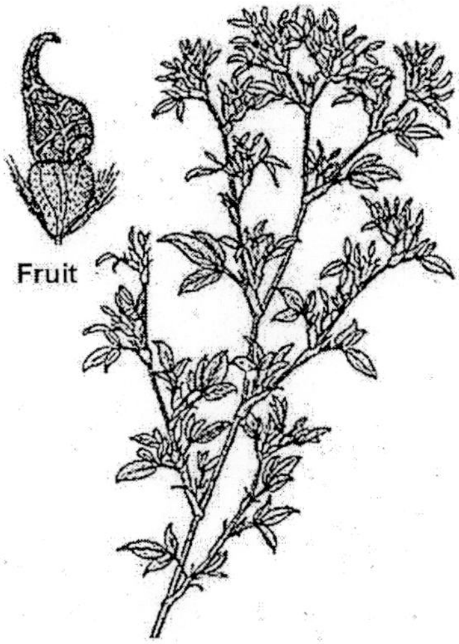

Fig. 9.2. Stylosanthes hamata twig with flowering and fruit with axis rudiment

Origin: West Africa & South America

Related species: Stylosanthus humilis & S. gracillis

Economic Importance:

- Suited for grazing as well as a fodder.
- It makes a good hay as it retains its nutritive value even when ripe.
- It is suited to grow under trees, Lauecana sp.

BOTANICAL FEATURES OF CLITORIA

Clitoria: *Clitoria ternatea* ($2n = 16$) **Family:** Fabaceae

Habitat	:	Tropical
Habit	:	A twining herbaceous perennials
Root	:	Tap root
Stem	:	Aerial, weak stem and a twiner
Leaves	:	Imparipinnately compound, alternate, stipulate showing reticulate venation. Leaflets are stipellate. Petiole and stipel are pulvinated
Inflorescence	:	Axillary raceme with single or paired flowers

Flower	:	Bracteate, bracteolate, bracteoles usually large, pedicellate, dichlamydeous, complete, bisexual, pentamerous, zygomorphic and hypogynous
Calyx	:	Five sepals, gamosepalous, tubular, valvate aestivation
Corolla	:	Five petals, polypetalous, vexillary aestivation
Androecium	:	Ten stamens, Monadelphous (9 + 1), Basifixed
Gynoecium	:	Monocarpellary, superior ovary
Fruit	:	Pod
Pollination	:	Self pollination

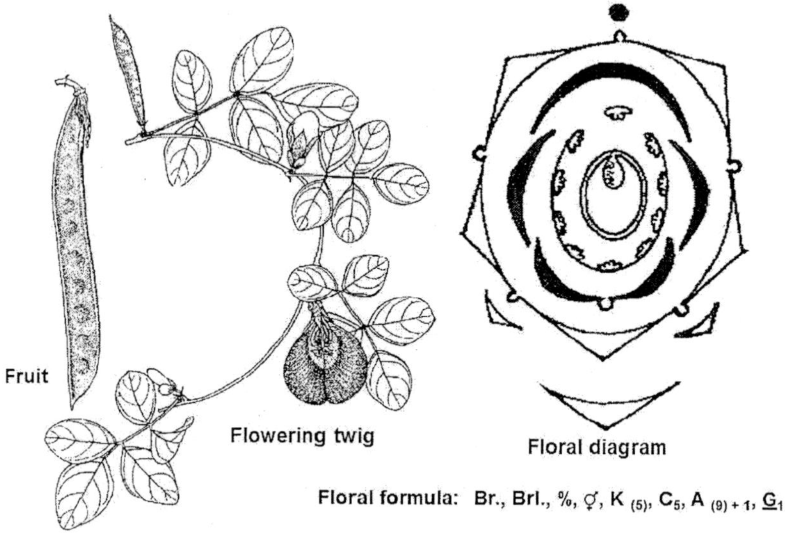

Fig. 9.3. Floral diagram and formula of Clitoria (Clitoria ternatea)

Origin: Tropical Africa

Related species: Clitoria laurifolia

Economic Importance:

It is grown in waste lands for fodder and field boundaries.

BOTANICAL FEATURES OF AGATHI

Agathi – *Sesbania grandiflora* (2*n* = 24) **Family:** Fabaceae

Habitat	:	Tropical
Habit	:	Perennial Shrub / Tree
Root	:	Tap root with nodules
Stem	:	Solid, cylindrical

Leaves	:	Paripinnately compound leaf, alternate, stipulate, petiolate, pulvinate, leaflets ovate, entire, acute, reticulate venation
Inflorescence	:	Axillary raceme
Flower	:	Pedicellate, bracteate, zygomorphic, bisexual, hypogynous, complete, pentamerous
Calyx	:	Five sepals, gamosepalous, valvate aestivation
Corolla	:	Five petals, polypetalous, vexillary aestivation
Androecium	:	Ten stamens, Diadelphous (9 + 1)
Gynoecium	:	Monocarpellary, Superior ovary and ovules in marginal placentation
Fruit	:	Pod; Seed: Non-endospermous, large size, thick, ovoid
Pollination	:	Self pollination

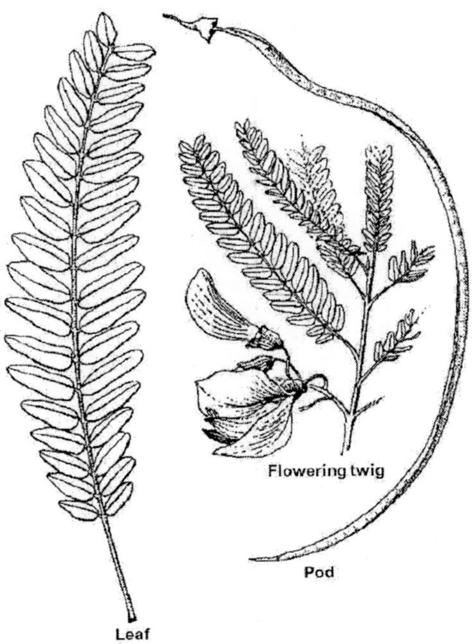

Fig. 9.4. Agathi: Sesbania grandiflora (Papilionaceae)

Economic Importance

- Used as fodder during dry season.
- Normally, grown in paddy field and it fixes nitrogen from admosphere.

BOTANICAL FEATURES OF SUNNHEMP

Sunnhemp (Channapai) – *Crotalaria juncea* ($2n = 24$) **Family:** Fabaceae

Horse gram is the poor man's pulse crop in Southern India. It is a hardy, drought resistant annual grain legume.

Habitat	:	Sub – tropical, Tropical
Habit	:	Tall, erect annual herb
Root	:	Tap root with nodules
Stem	:	Erect, ribbed, Branched, Pubescent
Leaves	:	Simple, stipulate, Opposite, petiolate, pulvinate, Spirally arranged, Lanceolate, entire, acute, reticulate venation
Inflorescence	:	Both terminal & Axillary raceme, flowers papilionaceous
Flower	:	Pedicellate, bracteates, zygomorphic, bisexual, hypogynous, complete, pentamerous
Calyx	:	Five sepals, gamosepalous, valvate aestivation, Bilabiate (2+3)
Corolla	:	Five petals, polypetalous, vexillary aestivation, bright yellow
Androecium	:	Ten stamens, Monadelphous, Didynamous (Long filament short anther and short filament long anther), Basifixed
Gynoecium	:	Monocarpellary, superior ovary
Fruit	:	Soft hairy pods
Pollination	:	Self pollination
Floral Formula &	:	As in Fig.9.5

Floral Diagram

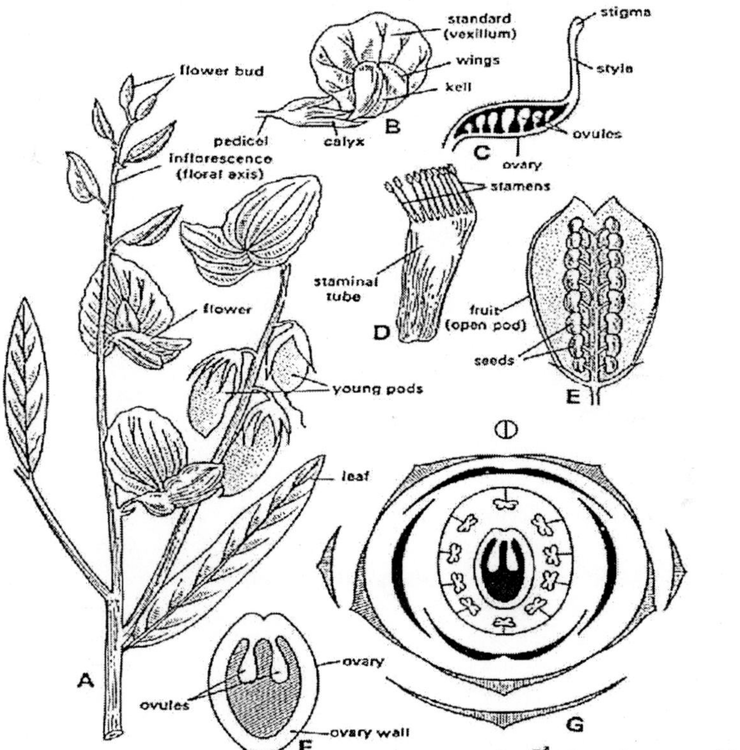

Br., Brl., %, ⚥, $K_{(5)}$, C_5, $A_{(10)}$, $\underline{G}_1$

A, Flowering twig with young pods; B, Flower; C, L.S. of Gynoecium; D, Androecium; E, Dehisced fruit; F, T.S. of Ovary; G, Floral diagram

Fig. 9.5. Sunnhemp: Crotalaria juncea (Papilionaceae)

Economic Importance: Bast fibre, one to 1.5 m long, lustrous and fairly resistant to micro organism. It is stronger than jute, lighter in colour.

Sunn hemp is a source of bast fibre, grown in tropical countries. It is also grown as a fodder and green manure. India exports this fibre to United Kingdom, Belgium and United States. India, Sri Lanka.

In India: Uttar Pradesh is the largest producer, other major producing areas are Bihar, Andhra Pradesh Tamil Nadu and Madhya Pradesh.

Place of origin: Asia

Uses: Cordage fibre, marine cordage, manufacture of sail cloth, canvas, matting, rope, soles of shoes and sandals; used for making cigarette papers, tissue paper, fishing net.

BOTANICAL FEATURES OF SESBANIA

(i) **Daincha** – ***Sesbania aculeata*** **(Family – Fabaceae):** A quick growing succulent crop which adapts itself to varying conditions of soil and climate. A soft stemmed shrub with or without prickles, growing to 1.5 -2.0 metres, leaves long, pinnately compound, with small oblong leaflets, stiuples small and deciduous, flowers yellowish in axillary racemes, partly drooping, fruiting profuse, fruit long, cylindrical with 25-30 seeds per pod, seeds oblong, greyish brown in colour. Nienty days crop yields upto 4500 kg of green material and a 150 days crop about 8000 Kg.

(ii) ***Sesbania speciosa*** **(Family – Fabaceae):** The plant has a grayish appearance with soft grayish tomentose hairs on stems and leaves. The stem is pithy but if allowed to grow it becomes woody.

Leaves compound, long with numerous leaflets, leaflets near the pulvinus the longest and gradually reduced in size towards the tip of the leaf. Stipules prominent with curved acuminate tip almost encircling the stem near the point of attachment. Inflorescence is a long lax drooping raceme with many yellowish flowers with purple splashes on the standard, wing petals deep yellow and the keels pale yellow, fruiting profuse, fruits, long, flat, quadrangular ending in a tapering tip.

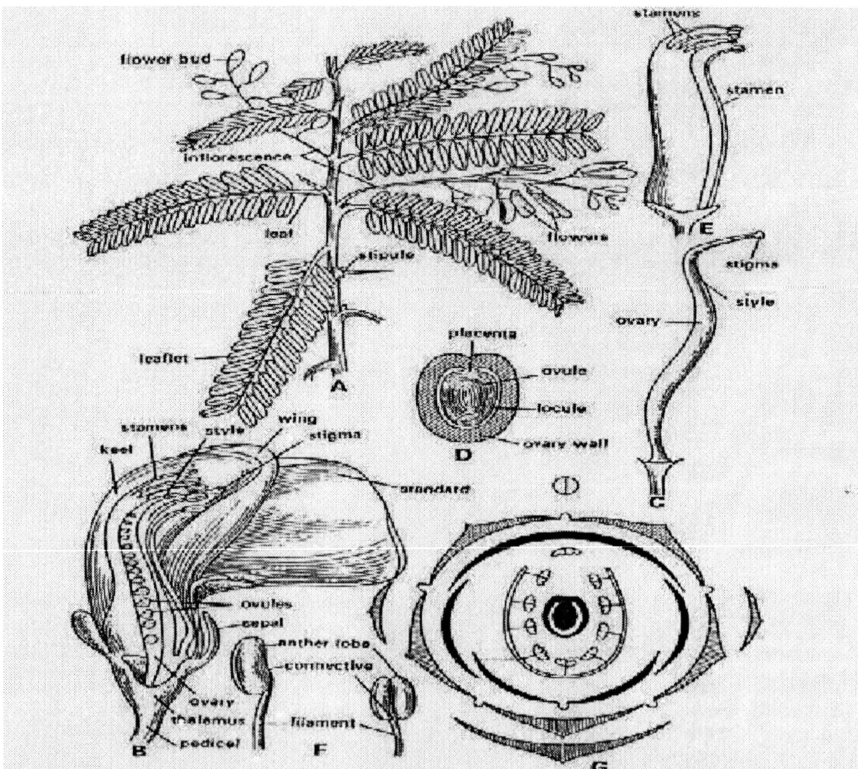

Fig. 15.10. Papilionaceae. *Sesbania sesban* (L.) Merr. (Jayanti). A, flowering twig; B, L.S., of flower; C, gynoecium; D, T.S. of ovary; E, androecium; F, stamens; G, floral diagram.

Fig 9.6. Sesbania aculeate floral biology

Chapter 10

List of Cultivated Crops, Economic parts, Chromosome Number and Family description of Pedaliaceae and Asteraceae : Key botanical features of Gingelly (Pedaliaceae), Sunflower, Safflower & Chrysanthemum (Asteraceae) and (Oleaceae) Jasmine

LIST OF CULTIVATED CROPS UNDER FAMILY PEDALIACEAE

Sl. No.	List of cultivated crops	Scientific name	Chromosome number	Economic Parts
	Oil seeds			
1	Gingelly (Sesame)	*Sesamum indicum*	26	Seed

Family description of Pedaliaceae (Sesame family)

Plant	:	Annual herb or under shrub with mucilageous glandular hairs
Stem	:	Erect, quadrangular and branched
Leaf	:	Simple, opposite at base & alternate on upper
Inflorescence	:	Axillary solitary cyme
Flower	:	Zygomorphic, bisexual, hypogynous
Calyx	:	Sepals – 5, Gamosepalous, Valvate
Corolla	:	Petals – 5, Gamopetalous, Bilabiate, Ventricose (bell shaped), Valvate
Androecium	:	Stamens – 4, Didynamous, Epipetalous
Gynoecium	:	Bicarpellary, Syncarpous, Superior overy, Bilocular due to formation of false septum, many ovules in axile placentation
Fruit	:	Capsule

Systematic position of family Pedaliaceae

Kingdom	:	Plant kingdom
Division	:	Phanerogams
Subdivision	:	Angiosperms
Class	:	Dicotyledonae
Subclass	:	Gamopetalae
Series	:	Bicarpellatae
Order	:	Personales
Family	:	Pedaliaceae

BOTANICAL FEATURES OF GINGELLY (*Sesame*)

- Gingelly is very ancient oil yielding crop which is rich in Vitamin E
- It is of great importance in medicine due to its insecticidal property

Gingelly (Sesame) *Sesamum indicum* $2n = 26$ **Family:** Pedaliaceae

Habitat	:	Tropical; Very sensitive to water logging
Habit	:	Erect annual herb
Root	:	Tap root system with a dense mat of feeding roots
Stem	:	Erect, quadrangular, longitudinally furrowed
Leaves	:	Simple, pedicellate, exstipulate, opposite below, spiral above, the leaves above are entire, lanceolate whereas the leaves below are broad & lobed
Inflorescence	:	Axillary solitary Cyme or in groups of two to three flowers
Flower	:	Pedicellate, Two aborted glands on either side of the pedicel, Ebractiate, Zygomorphic, Bisexual, Complete, Hypogynous, Pinkish in colour
Calyx	:	Sepals – 5, Gamosepalous, Tubular, Valvate aestivation
Corolla	:	Petals – 5, Gamopetalous, Ventricose (bell shaped), Two lipped (Bilabiate): Upper lip consists of two equal sized lobes while lower lip includes three petals of which the central one is the largest, Hairy outside, a ring of hairs is often present inside the corolla towards the base below the stamens.
Androecium	:	Stamens 4, small posterior staminode represents fifth stamen, Epipetalous, Didynamous
Gynoecium	:	Bicarpellary, Syncarpous, Superior ovary, Bilocular due to formation of false septa, many ovules in axile placentation. The style is forked at the tip and carries the two stigmatic hairy lobes

Fruit	:	Capsule, loculicidal dehiscence, two types of fruits are recognized. One is four loculed and the other eight loculed
Pollination	:	Often cross pollination
Floral Formula &	:	As in Fig. 10.1
Floral Diagram	:	

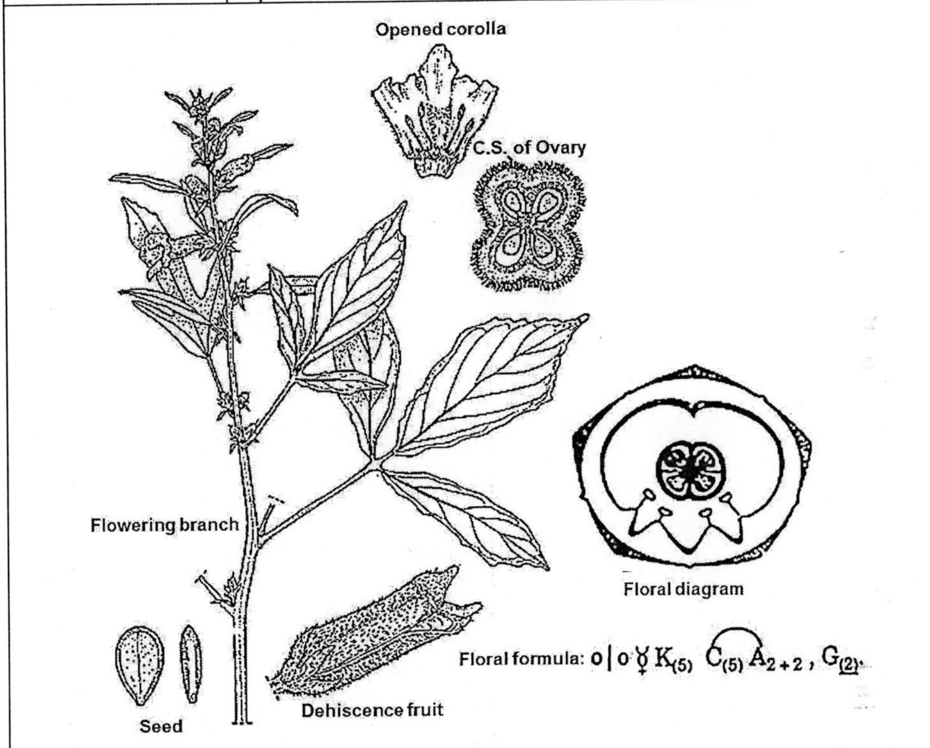

Fig. 10.1. Floral structure and diagram of Sesamum Indicum (Pedaliaceae)

Origin: Africa

Putative parent:

Diploids		Tetraploids	
S. prostratum	2n = 32	*S. radiatum*	2n = 64
S. laciniatum	2n = 32	*S. occidentale*	2n = 64
S. alatum	2n = 26		
S. grandiflorum	2n = 26		

Economic Importance:

- Seeds mixed with jiggery and eaten – chikki.
- Good source of cooking oil (85% unsaturated fatty acid).

- Manufacture of margarine.
- Soap, paint, illuminant, base for scented oil
- Carrier for antibiotics, vitamins and hormones.
- Oil cake – rich in calcium, phosphorus and the vitamin niacin : used as cattle feed.
- Seed and green plant have medicinal value.

List of cultivated crops under family Asteraceae / Compositae

Sl. No.	List of cultivated crops	Scientific name	Chromosome number	Economic Parts
	Oil seeds			
1	Sunflower	*Helianthus annuus*	34	Seed
2	Safflower	*Carthamus tinctorius*	24	Seed
	Ornamental Flower			
1	Chrysanthemum	*Chrysanthemum indicum*	18	Flower

Family description of Asteraceae / Compositae

Plant	:	Annual / Perennial herb
Stem	:	Erect, Solid, Hairy
Leaf	:	Simple, Exstipulate, Alternate, Ovate / Cardate, Dentate leaf margin, Acute apex, Reticulate venation
Inflorescence	:	Head / Capitulam surrounded by an involucre of bracts
Flower	:	Florets are of two types, 1. Outer unisexual femalc, 2. Inner bisexual
Calyx	:	Calyx modified into numerous scales called pappus
Corolla	:	Petals – 5, gamopetalous, valvate aestivation
Androecium	:	Five stamens, Syngenesious, Epipetalous, Basifixed
Gynoecium	:	Carpels -2, Syncarpous, Inferior ovary, Single ovule in basal placentation, Bifid stigma
Fruit	:	Cypsella / Achene

Systematic position of family Asteraceae

Kingdom	:	Plant kingdom
Division	:	Phanerogams
Subdivision	:	Angiosperms
Class	:	Dicotyledonae

Subclass : Gamopetalae
Series : Inferae
Order : Asterales
Family : Asteraceae

BOTANICAL FEATURES OF SUNFLOWER

- Sunflower is an important oilseed crop in the world after soybean and oilpalm
- Its oil content is ranging from 35 to 45 per cent and it has non-cholesterol and anti cholesterol property

Sunflower: *Helianthus annuus* ($2n = 34$) **Family:** Asteraceae

Habitat	:	Tropical and subtropical
Habit	:	Tall annual herb. Two varieties: Tall unbranched, single headed type Short, branched, multi-headed type
Root	:	Tap root with large lateral spread of surface roots
Stem	:	Erect, rough, hairy and pithy
Leaves	:	Simple, Petiolate, Exstipulate, Lower leaves – opposite decussate, Upper leaves - alternate, Ovate / Cordate shape, Dentate leaf margin, Acute apex, Reticulate venation, Hairs on both sides
Inflorescence	:	Heterogamous Capitulum (Head), Terminal. Receptacle is flat, slightly convex with two to three rows of large pointed involucres of bracts
Flower	:	Head has outer row of ray florets and inner rows of disc florets. Florets are subtended by broad scales (Bracts). Ray florets Found in the peripheral region of head. Sessile, Bracteate, Zygomorphic, Incomplete, Unisexual, Pistillate, Pentamerous, Epigynous
Calyx	:	Sepals reduced into numerous scales called pappus
Corolla	:	Petals – 5, Gamopetalous, Valvate, Ligulate, 0/5
Androecium	:	Stamens – 5, Syngenesious, Epipetalous, Basifixed
Gynoecium	:	Carpels – 2, Syncarpous, Inferior ovary, Single ovule in basal placentation, Short style, Bifid stigma
Fruit	:	Cypsella
Floral Formula	:	As in 10.2

Floral Diagram :

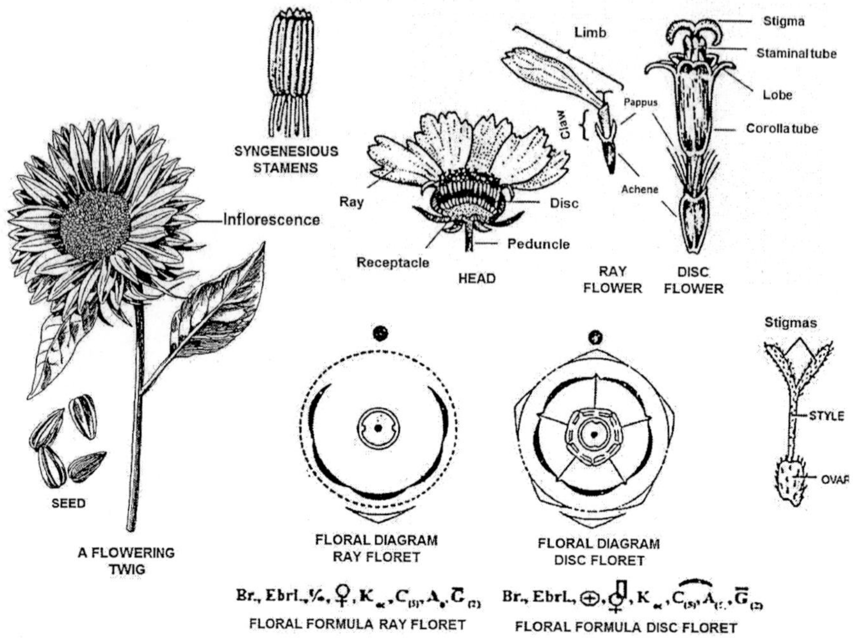

Fig. 10.2. Floral diagram and formula of Helianthus annuus(Asteraceae)

		Disc florets Found in the center of head. Sessile, Bracteate, Actinomorphic, Complete, Bisexual, Pentamerous, Epigynous
Pollination	:	Cross pollination

Origin: North America

Wild species: *H. hirsutus, H. rigidis, H. petiolaris*

Cultivated species:

H. annuus - Sunflower, Diploid (2n=34) – Cultivated for its oil

H. tuberosus - Jerusalem Artichoke, Hexaploid (2n=102) - Cultivated for its tubers

Economic Importance:

- Oil is used as cooking media (90% poly unsaturated fatty acid (PUFA) and 10% Saturated fatty acid).
- Oil has non cholesterol and anti-cholesterol properties used in cooking and for salads.
- Oil is used in paints, varnishes, soap, cosmetics.

- Oil cake and hulls are used as cattle feed and fuel. From dried stalk paper is made.
- Sunflower also produces excellent honey and wax.
- Fried sunflower seeds are edible.

BOTANICAL FEATURES OF SAFFLOWER

- Safflower is an important oil seed crop in India
- It is cultivated for both oil and orange dye (Carthamin)

Safflower: *Carthamus tinctorius* ($2n = 24$) **Family:** Asteraceae

Habitat	:	Tropical and Subtropical, Grown as rainfed crop
Habit	:	Annual herb, much branched, plants may be glabrous or pubescent
Root	:	Tap root, long and stout
Stem	:	Erect, woody, glabrous and grooved
Leaves	:	Simple, Sessile, Spirally arranged, Lanceolate leaf shape, Spinulose serrated leaf margin, reticulate venation
Inflorescence	:	Homogamous Capitilum (Head), Terminal. Spiny bracts forming involucre
Flower	:	Bracteate (modified into thin soft hairs), Actinomorphic, Bisexual, Tubular, Orange / yellow coloured, Epigynous
Calyx	:	Sepals reduced into numerous scales called pappus
Corolla	:	Petals -5, Gamopetalous, Long tubular,
Androecium	:	Stamens – 5, Syngenesious, Epipetalous
Gynoecium	:	Carpels – 2, Syncarpous, Inferior ovary with basal placentation, Bifid stigma
Fruit	:	An achene which is obovate, exendospermous, Quadrangled
Pollination	:	Often cross pollination
Floral Formula &	:	As in fig 10.3

Floral Diagram :

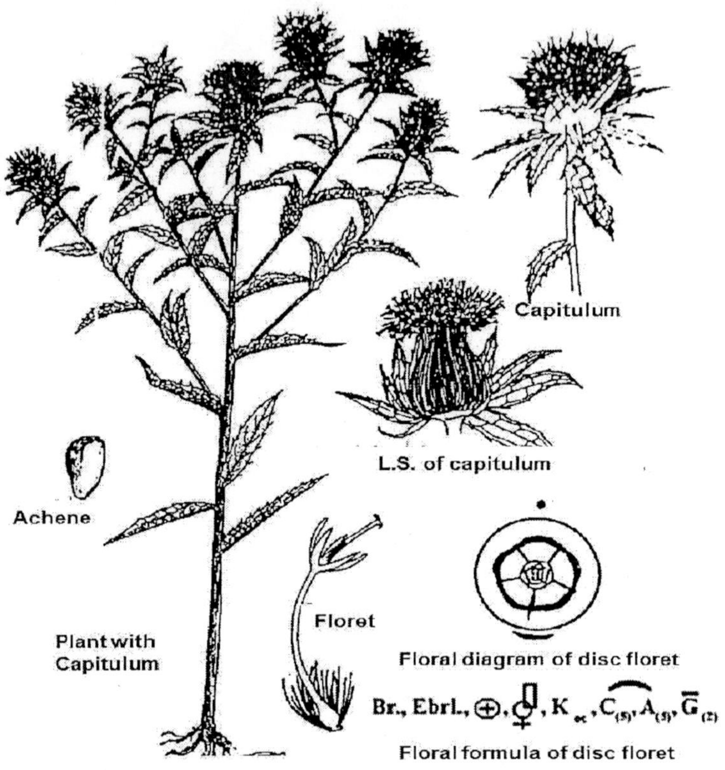

Fig.10.3. Floral diagram and formula of Carthamus tinctorius (Asteraceae)

Origin: Afghanistan,

Wild species: *Carthamus lunatus (Progenitor), Carthamus oxyacantha*

Economic Importance:

- Cooking media (Saffola)
- Good drying oil
- Corolla used in dyeing industry
- Oil cake – stock feed
- Preparation of paints and varnishes.

BOTANICAL FEATURES OF CHRYSANTHEMUM

- It was first cultivated in China as flowering herb.
- Greek word 'Chrysous' means the golden colour of the flower.

Chrysanthemum – *Chrysanthemum indicum* ($2n = 18$) **Family:** Asteraceae

Habitat	:	Tropical and Subtropical
Habit	:	Perennial herb
Root	:	Tap root
Stem	:	Erect, woody, branched
Leaves	:	Pedicellate, Alternate, Deeply lobed, elongated and toothed
Inflorescence	:	Heterogamous Capitilum (Head), Terminal.
Flower	:	Heads many flowered; disc florets perfect and mostly fertile; ray florets pistillate, mostly fertile; receptacle flat or convex, Pappus absent
Pollination	:	Often cross pollination

Center of origin: China

Related wild species: Chrysanthemum sagetum

Economic importance:

- Chrysanthemum is used as cut flowers, as adorne to God
- Used as landscape bedding plants
- Pyrethrum, a natural insecticide derived from Chrysanthemum cinerariaefolium
- Extracts of stem and flower has anti-HIV and anti bacterial property
- Plant extract can be applied on skin infections

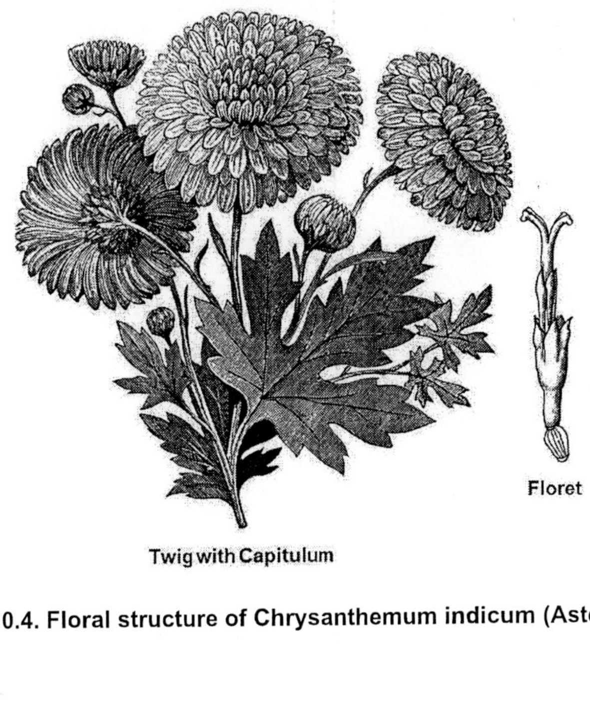

Fig. 10.4. Floral structure of Chrysanthemum indicum (Asteraceae)

List of cultivated crops under family Oleaceae

Sl. No.	List of cultivated crops	Scientific name	Chromosome number	Economic Parts
1	Jasmine	Jasminum officinale	26 & 52	Flower

Family description of Oleaceae

Plant	:	Plants are Climbers, Shrubs or Trees
Stem	:	Woody / twin
Leaf	:	Simple / Compound (Pinnate or ternate), Opposite
Inflorescence	:	A simple dichasial Cyme
Flower	:	Actinomorphic, Bisexual
Calyx	:	May or may not be present
Corolla	:	Petals – 5, Gamopetalous
Androecium	:	Stamens – 2, inserted in perigynous zone
Gynoecium	:	Bicarpellary, Bilocular
Fruit	:	Berry

Systematic position of family Oleaceae

Kingdom	:	Plant kingdom
Division	:	Phanerogams
Subdivision	:	Angiosperms
Class	:	Dicotyledonae
Subclass	:	Gamopetalae
Series	:	Bicarpellatae
Order	:	Lamiales
Family	:	Oleaceae

BOTANICAL FEATURES OF JASMINE

Jasmine: *Jasminum officinale* $2n = 26$ & 52 **Family:** Oleaceae

Habitat	:	Tropical and Subtropical
Habit	:	Semi vining shrub
Root	:	Tap root
Stem	:	Slender trailing and glabrous
Leaves	:	Imparipinnately compound with 5 – 9 leaflets, leaflets entire
Inflorescence	:	A simple dichasial Cyme
Flower	:	Actinomorphic, Bisexual generally white in colour

Calyx	:	Sepals – 5, Gamosepalous, Tubular
Corolla	:	Petals -5, Gamopetalous, tubular
Androecium	:	Stamens – 2, Polyandrous
Gynoecium	:	Carpels – 2, Bilocular
Fruit	:	Berry

Other species of Jasmine

Jasminum sambac	:	Arabian Jasmine (Gundu malli)
Jasminum auriculatum	:	Mullai
Jasminum grandiflorum	:	Spanish Jasmine (Jathimalli, Pitchi)
Jasminum polyanthum	:	Santhana mullai, Thalavam
Jasminum officinale	:	Poet's Jasmine (Mauval)
Jasminum pubescens	:	Indian Jasmine (Kokudi)
Jasminum angustifoliam	:	Wild Jasmine (Kattu malli)

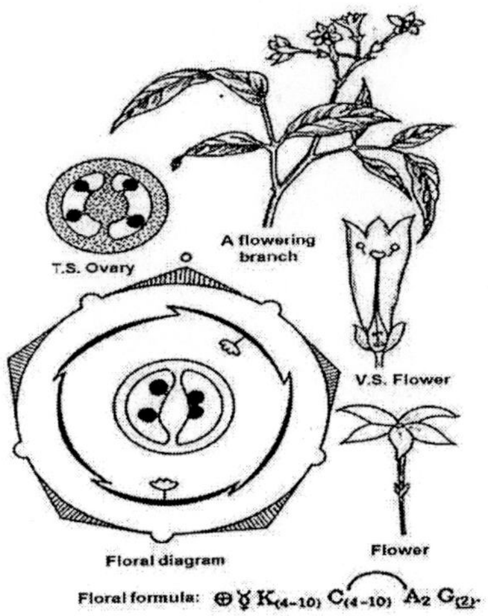

Fig. 10.5. Floral diagram and formula of Jasminum officinale (Oleaceae)

Chapter

11

List of cultivated crops, economic parts, chromosome number and family description of Brassicaceae and Euphorbiaceae; Key botanical features of Rape seed, Mustard, Cabbage & Cauliflower (Brassicaceae) and Castor, Jatropha & Tapioca (Euphorbiaceae)

LIST OF CULTIVATED CROPS UNDER FAMILY BRASSICACEAE (*Cruciferae*)

Sl. No.	List of cultivated crops	Scientific name	Chromosome number	Economic Parts
Rape & Mustard				
1	Rape seed	*Brassica napus*	38	Seed
2	White Mustard	*Brassica alba*	24	Seed
3	Field Mustard	*Brassica campestris*	20	Seed
4	Indian Mustard	*Brassica juncea*	36	Seed
5	Black Mustard	*Brassica nigra*	16	Seed
Vegetables				
1	Cabbage	*Brassica oleraceae var. capitata*	18	Apical Bud
2	Cauliflower	*Brassica oleraceae var. botrytis*	18	Aborted floral meristem

Family description of Brassicaceae (Cruciferae)

Plant	:	Annual herb with pungent watery latex
Stem	:	Erect, Cylindrical, Soild, Branched, Glabrous
Leaf	:	Simple, Alternate, Exstipulate, Lyrate, Glabrous, Reticulate
Inflorescence	:	A raceme
Flower	:	Pedicellate, Ebracteate, Actinomorphic, Bisexual, Complete, Tetramerous, Hypogynous

Calyx	:	Sepals – 4 in two whorls, Polysepalous
Corolla	:	Petals – 4, Polypetalous, Cruciform, Each petal is distinguished into a claw and a limb, Valvate aestivation.
Androecium	:	Stamen – 6, Tetradynamous, Polyandrous, Basifixed
Gynoecium	:	Carpels – 2, Syncarpous, Superior overy, unilocular but later becomes bilocular due to false septum (replum)
Fruit	:	Siliqua

Systematic position of family Brassicaceae (Cruciferae)

Kingdom	:	Plant kingdom
Division	:	Phanerogams
Subdivision	:	Angiosperms
Class	:	Dicotyledonae
Subclass	:	Polypetalae
Series	:	Thalamiflorae
Order	:	Parietales
Family	:	Brassicaceae

RAPE SEED / MUSTARD

The group rape and mustard includes the oil yielding species of Brassica

Rape seed / Mustard: *Brassica sp.* $2n = 16,18,20,22$ & 36

Family: Brassicaceae

Rape seed

1. Indian Rape seed: *Brassica campestris* ($2n = 20$)
   Cultivated types: *Brassica campestris var. brown sarson*
   Brassica campestris var. yellow sarson
   Brassica campestris var. toria
2. European Rape seed: *Brassica napus* ($2n = 38$)

Mustard

1. *Brassica nigra* (2n = 16): Black mustard
2. *Brassica alba* (2n = 24): White mustard
3. *Brassica juncea* (2n = 36): Indian mustard

Genetic relationship of Brassica sp. – Triangle of 'U'

B. nigra
n = 8
BB

B. carinata
n = 17
BC

B. juncea
n = 18
AB

B. oleraceae
n = 9
CC

B. napus
n = 19
AC

B. campestris
n = 10
AA

BOTANICAL FEATURES OF MUSTARD

Habitat	:	Low humid tropics
Habit	:	Annual herb with pungent watery latex
Root	:	Tap root
Stem	:	Erect, Cylindrical, Soild, Branched, Smooth
Leaves	:	Simple, Alternate, Exstipulate, Lyrate, Glabrous, Reticulate venation
Inflorescence	:	A raceme
Flower	:	Pedicellate, Ebracteate, Actinomorphic, Bisexual, Complete, Tetramerous, Hypogynous, Small yellow coloured
Calyx	:	Sepals – 4 in two whorls, Polysepalous, free imbricate
Corolla	:	Petals – 4, Polypetalous, Cruciform, Each petal is distinguished into a claw and a limb, Valvate aestivation.
Androecium	:	Stamen – 6, Tetradynamous, 4 inner long & 2 outer short, Polyandrous, Basifixed, glands are present at the base of 4 longer stamens
Gynoecium	:	Carpels – 2, Syncarpous, Superior overy, unilocular but later becomes bilocular due to false septum (replum), Many ovules in parietal placentation, Capitate stigma
Fruit	:	Siliqua
Pollination	:	Cross pollination

Floral Formula	:	Floral Formula. 4 sepals, 4 petals, 6 stamens, and two carpels. The carpels form a superior ovary with false septum and two rows of campylotropous ovules. The androecium is tetradynamous, i.e. there are two short and four long stamens. The sepals are erect. The nectar is secreted by two nectaries situated between the basis of the short stamens and the ovary.
Floral Diagram	:	As in fig.11.1

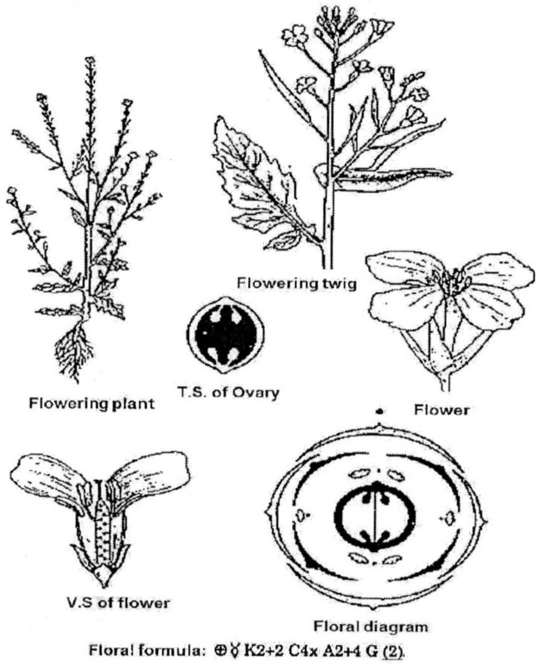

Fig. 11.1. Floral structure and diagram of Brassica cambestris (Brassicaceae)

Origin: India

Economic Importance:

- Seed as condiment
- Oil as cooking oil
- Oil – illuminant, lubricant, soap industry, fertilizer industry, plastic industry.
- Green plant as fodder. Essential oil in mustard has medicinal properties. Used as counter irritant. The leaves of young plants used as green / leafy vegetable.

BOTANICAL FEATURES OF CABBAGE

It is leafy green vegetable distinguished by a short stem upon which a mass of leaves form. As the plants grow leaves increase in number forming a head in the center of the plant

bbage: *Brassica oleraceae var. capitata* $2n = 18$ **Family:** Brassicaceae

Habitat	:	It is cool season crop tolerates cold and frost
Habit	:	Biennial herb with short stem upon which globular cabbage head forms
Root	:	Tap root
Stem	:	Short stem with cluster of compact leaves
Leaves	:	Cluster of leaves form a compact globular head. Leaves are smooth and crinkly textured. Leaves are broad and cup shapes with reticulate venation

Economical importance

- Cabbage head id used as vegetable.
- It can be consumed raw in the form of salad.
- Cabbage is rich source of beta-carotein, Vitamin C and fibre.

Cabbage head

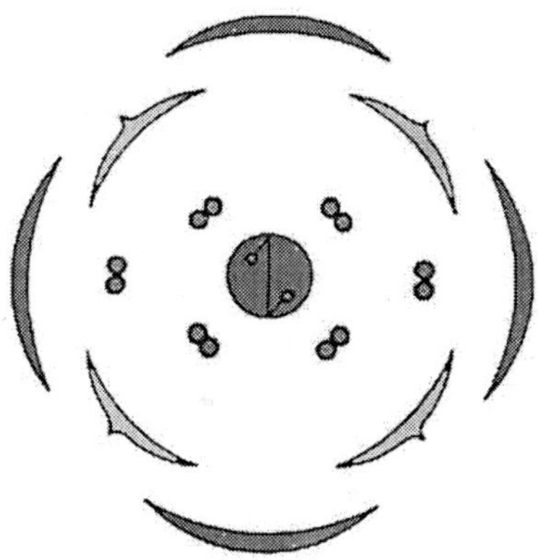

Floral diagram

Fig. 11.2. Cabbage: Brassica oleraceae var. capitata

BOTANICAL FEATURES OF CAULIFLOWER

Cauliflower: *Brassica oleraceae var. botrytis* ($2n = 18$) **Family:** Brassicaceae

Cauliflower can be annual or biennial crops. Cauliflower has a pre-floral structure that possesses characteristics of both a vegetative and reproductive apex. Morphologically, it is an early stage of inflorescence development, since its formation takes place before floral initiation. Inflorescence is a curd with tightly branched peduncles containing thousands of meristems at an early floral stage.

The inflorescense is a raceme, and flowering starts from the bottom to the top of the inflorescense. Cauliflower, has yellow flowers with 4 petals, 4 sepals, 6 stamens, one pistil containing two carpels with 10 to 15 ovules inside each carpel.

After fertilization of the ovules, seeds form inside a pod called a silique. This structure is highly dehiscent. Thus, harvest must be done before the siliques shatter. Seeds are round and reddish brown to black in color. The seed is cleaned on an air-screen cleaner. A spiral separator is used to separate splits from whole seeds.

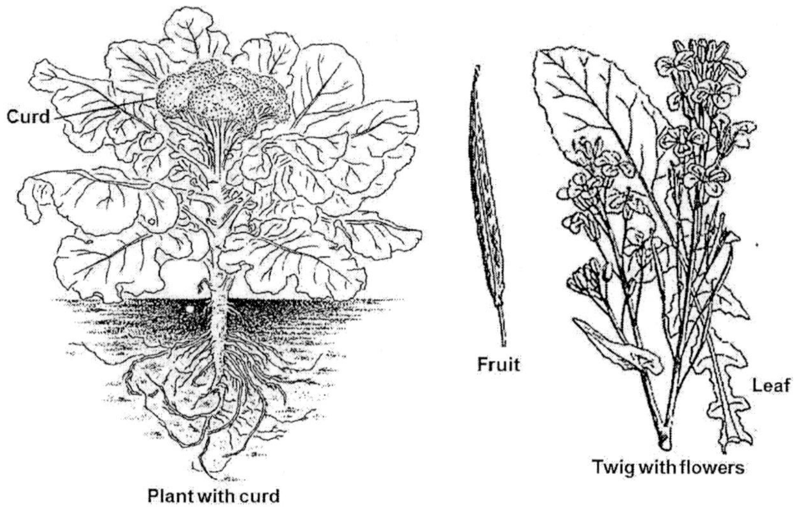

Fig. 11.3. Cauliflower: Brassica oleraceae var. botrytis (Brassicaseae)

List of cultivated crops under family Euphorbiaceae

Sl. No.	List of cultivated crops	Scientific name	Chromosome number	Economic Parts
	Oil seed Crops			
1	Castor	*Ricinus communis*	20	Seed
2	Jatropha	*Jatropha curcas*	22	Seed
	Starch Crop			
1	Tapioca	*Manihot esculenta*	36	Root tuber

Family description of Euphorbiaceae

Plant	:	Annuals with latex, Monoecious
Stem	:	Erect, Solid, Lower portion woody, Glabrous, Branched milky latex present, Green to purple coloured
Leaf	:	Simple, Petiolate, Alternate, Stipulate, reticulate venation
Inflorescence	:	Terminal panicle with unisexual flowers arranged in racemose manner
Flower	:	Bracteate, Pedicellate, Unisexual, Incomplete and Hypogynous
Perianth	:	Perianth usually in one whorl or absent

Androecium	:	Stamen is polyadelphous and the filaments are branched. They are fused into several bundles
Gynoecium	:	Tricarpellary, Syncarpous, Ovary superior, Ovary trilocular with one or two ovules in each locule on axile placentation. Ovary is distinctly three lobed. Styles three, each ending in a bifid stigma.
Fruit	:	Regma, dehiscing into three cocci

Systematic position of family Euphorbiaceae

Kingdom : Plant kingdom
Division : Phanerogams
Subdivision : Angiosperms
Class : Dicotyledonae
Subclass : Monochlamydeae
Series : Unisexuales
Family : Euphorbiaceae

BOTANICAL FEATURES OF CASTOR

- Castor is an important commercial crop and industrially valued for the oil
- Brazil is the largest producer of castor followed by India, USSR & Argentina

Castor: *Ricinus communis* ($2n = 20$) **Family:** Euphorbiaceae

Habitat	:	Temperate & Subtropical
Habit	:	Annual herb / Shrub with latex, Monoecious
Root	:	Tap root
Stem	:	Erect, Solid, Lower portion woody, Glabrous, Branched, milky latex present, Green to purple coloured
Leaves	:	Simple, Petiolate, Alternate, Stipulate, Palmately lobed with serrated margin, Acute apex, glabrous, milky latex present, Palmately reticulate venation
Inflorescence	:	Terminal panicle with unisexual flowers, Male flowers at the base & female flowers at the apex
Flower	:	Male flower: Pedicellate, Bractiate, Actinomorphic, Incomplete, Unisexual, Staminate, Pentamerous
Perianth	:	Tepals – 5, Gamophyllous, Valvate aestivation
Androecium	:	Stamens many, polyadelphous, filaments branched and united to form five branches. Anthers are globose, basifixed
Gynoecium	:	Tricarpellary, Syncarpous, Superior ovary, Trilocular with single ovule in each locule in axile placentation, Style – Absent, Stigma – 3 & Bifid
Floral Formula &	:	As in Fig 11.4

Floral Diagram

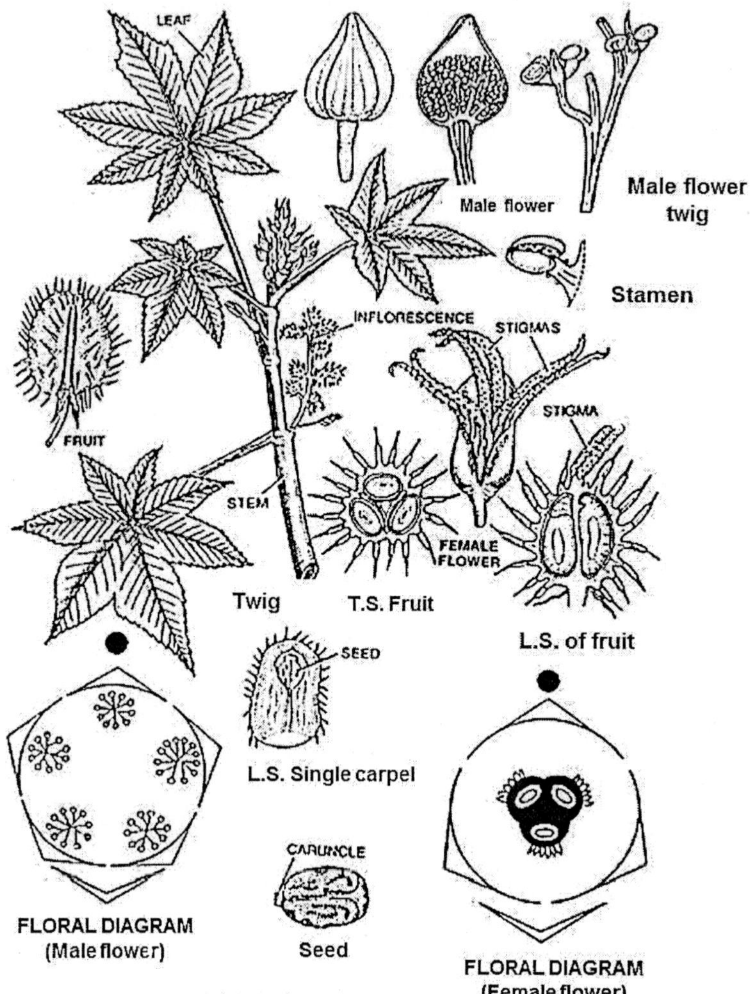

Floral Formula :

Male: ⊕ ♂ K 5, C 0, A 5 (branched)

Female: ⊕ ♀ K 3, C 0, G $(\underline{3})$.

Fig. 11.4. Floral diagram and formula of Ricinus communis (Euphorbiaceae)

Flower	:	Female flower: Pedicellate, Bracteate, Actinomorphic, Unisexual, Pistillate, Trimerous, Hypogynous
Perianth	:	Tepals – 3, Gamophyllous, Valvate aestivation
Fruit	:	Regma splitting into three cocci
Pollination	:	Cross pollination due to protogynous

Origin: Africa

Related species:

Ricinus zanzibarensis

Ricinus gibsoni

Ricinus sanguineus

Ricinus chinensis

Economic Importance:

- Used in paints, varnishes and other protective coverings.
- Used as illuminant.
- Used as lubricant for aeroengines.
- Used as hydraulic brake fluid.
- Used in manufacture of scap, printing ink, wax, polish.
- Used in medicine mainly as purgative.

BOTANICAL FEATURES OF JATROPHA

Jatropha: *Jatropha curcas* ($2n = 22$) **Family:** Euphorbiaceae

Jatropha is also called purging nut (English), Ratanjyot (Hindi) and kampma ramda (Sanskrit). It is a perennial deciduous shrub growing up to 3-5 metres height, shedding leaves during the dry season.

Origin

Jatropha is native to Mexico and central America, but is widely distributed in wild and semi-cultivated forms in Latin America, Africa, India, South-east Asia. In India, it was introduced in the 16th century by Portuguese naviagators.

Botany

The plant is deciduoud shrub and monoecious; Leaves are alternate, exstipulate, broad ovate / cardate; Inflorescence is called loose panicle; Flowers are unisexual. Fruit is known as capsule. Seeds resemble to castor seeds.

Economic importance

- Jatropha is grown for its seeds that are rich in non-edible oil (27-30%). The oil can be combusted as fuel without refining. The oil can be blended up to 20% with diesel to run diesel engines.
- The deoiled cake serves as organic manure.
- Besides oil, other plant parts have immense medicinal and insecticidal roles.
- The leaves are used for cure of cough, dyspepsia, leprosy and gynaec diseases.

The tender twigs provide relief to rheumatism. The green leaf manuring provides not only nutrients to crops but also provide effective control of Meloidogyne javonica nematode. The dried leaves can be used in storing food grains, as they deter various storage pests similar to neem leaves.

- The seed paste is used for diseases like snake bite, paralysis, piles, joint pains etc.
- The flowers are effective against diabetics.
- The root and bark have insecticidal properties.
- In Philippines, thread ropes are preferred from the bark.
- The deep root system enables it to serve as a soil binder, and thus helps in stabilization of sand dunes in deserts.
- It also grown as live fence/hedge for protection of farmer's field against damage by stray animals and blue bulls.

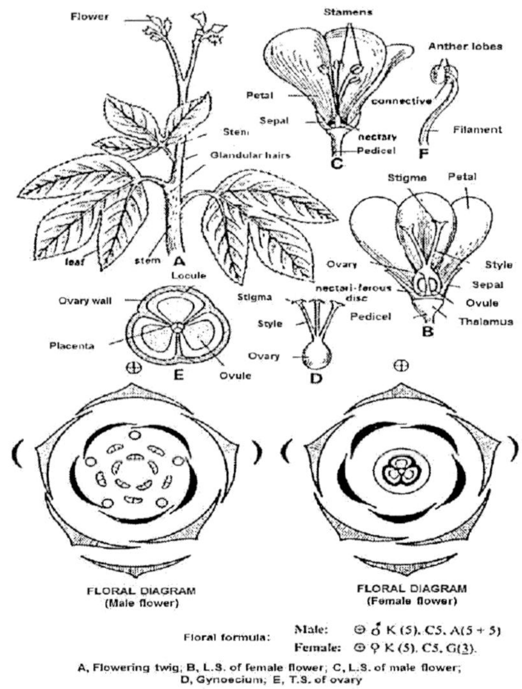

Fig. 11.5. Floral diagram and formula of Jatropha spp. (Euphorbiaceae)

BOTANICAL FEATURES OF TAPIOCA (Cassava)

Tapioca: *Manihot esculenta* (2*n* = 36) **Family:** Euphorbiaceae

Habitat	:	Tropical
Habit	:	Slender shrub, 2-7 m high, sparingly branched
Root	:	Tap root with tubers

Stem	:	Thick with prominent leaf scars, variously coloured
Leaves	:	Simple, Petiolate (long stalked), Alternate, palmately lobed, glabrous
Inflorescence	:	Axilary racemes, monoecious with male flowers at the top and female flowers in the bottom of inflorescence
Flower	:	Each flower (male and female) have five united sepals which may be yellow tinged. There are no petals (apetalous). Male flower has ten stamens arranged in two whorls of five stamens each. The filaments are free and anthers small. Female flower has an ovary with three locules and six ridges. Each locule contains a single ovule. The stigma has three lobes which unite to form a single style.
Floral formula & diagram	:	As in fig. fig.11.6
Fruit	:	Capsule with longitudinal wings
Pollination	:	Cross pollination due to protogyny

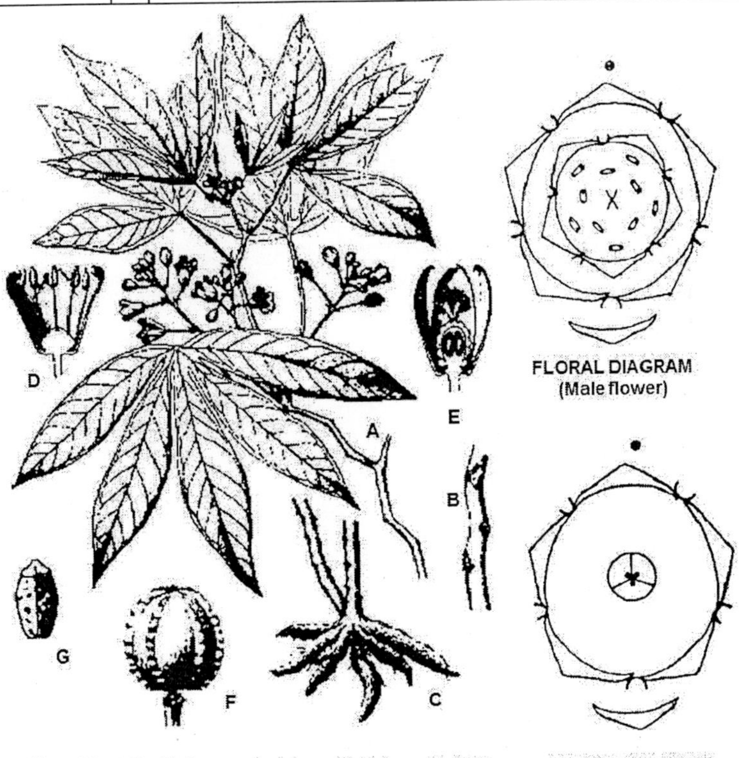

A, Flowering twig; B, Segment of stem; C, Tubers; D, Male flower; E, Female flower; F, Fruit; G, Seed

Fig.11.6. Floral structure and diagram of Manihot esculenta(Euphorbiaceae)

Place of origin: Central America.

Cultivated species: There are no wild species seen in the cultivated Manihot esculenta. The cultivated cassava can be classified into two broad groups as below

(a) **Sweet cassava:** Shorter in duration ,tubers maturing in 6-9 months. The cyanogenic glucoside is confined mainly to the outer skin (periderm).

(b) **Bitter cassava:** Longer in duration 12-18 months to mature, the cyanogenic glucoside is distributed throughout the tuber including core. The glycoside will be more in varieties having yellow flesh. Bitter cassava is suitable for cultivation where there is menace by wild boar is more.

Economic importance:

- Cooked tapioca tuber is popular food in Kerala normally taken with fish curry.
- Roots of tapioca is used to prepare chips.
- Used as source of starch in many sago industries.

Chapter

12

List of cultivated Crops, Economic parts, Chromosome Number and Family description of Arecaceae and Malvaceae; Key Botanical features of Coconut, Arecanut, Oil palm & Sugar palm (Arecaceae) and Cotton, Mesta & Bhendi (Malvaceae)

LIST OF CULTIVATED CROPS UNDER FAMILY ARECACEAE / PALMAE

Sl. No.	List of cultivated crops	Scientific name	Chromosome number	Economic Parts
	Oil seeds crop			
1	Coconut	*Cocos nucifera*	32	Endosperm
2	Oil palm	*Elaeis guineensis*	32	Mesocarp & Endosperm
	Masticatories			
1	Arecanut	*Areca catechu*	32	Nut
	Sugar crop			
1	Sugar palm	*Borassus flabellifer*	32	Inflorescence

FAMILY DESCRIPTION OF ARECACEAE / PALMAE

Plant	:	Perennial Trees
Stem	:	Stout, unbranched stem with persistent leaf base
Leaf	:	Very large pinnatisect / palmatisect compound leaves forming a crown at the top of the tree
Inflorescence	:	Spadix
Flower	:	Sessile, Actinomorphic, Unisexual
Perianth	:	Six in two whorls and sepaloid
Androecium	:	Stamens - 3+3
Gynoecium	:	Tricarpellary, Syncarpous, Superior ovary, Unilocular with one ovule in each locule
Fruit	:	Berry / Drupe

Systematic position of family Arecaceae / Palmae

Kingdom	:	Plant kingdom
Division	:	Phanerogams
Subdivision	:	Angiosperms
Class	:	Monocotyledonae
Series	:	Calycineae
Family	:	Arecaceae (palmae)

BOTANICAL FEATURES OF COCONUT

India is the third largest coconut producing country in the world next to Philippines and Indonesia. The coconut palm is a versatile crop.

Coconut: *Cocos nucifera* (2*n*=32) **Family:** Arecaceae / Palmae

Habitat	:	Tropical and subtropical
Habit	:	Perennial, Tall unbranched tree with terminal crown
Root	:	Mass of adventitious roots
Stem	:	Stout, Erect and unbranched with scars of leaf base
Leaves	:	Pinnately compound leaves, Petiolate, Exstipulate, Spirally arranged on the top of the trunk, Leaf segment: Sessile, Linear, Entire, Cuspidate, Parallel venation
Inflorescence	:	Axillary composite spadix consist of numerous drooping spikes enclosed in a spathe
Flower	:	Sessile, Bracteate, Actinomorphic, Unisexual, Hypogynous Male flowers: borne individually or in two or three, Upper portion of spike consists of male flowers
Perianth	:	Tepals - Six, arranged in two whorls. Three outer tapels are smaller than the inner three, Polyphyllous, Valvate aestivation
Androecium	:	Stamens - six, in two whorls, Basifixed
Gynoecium	:	Ovary large tricarpellary, Syncarpous, three ovules, two will abort and one will develop, Three sessile stigma
Floral Formula	:	As in Fig 12.1
Floral Diagram	:	As in Fig 12.1

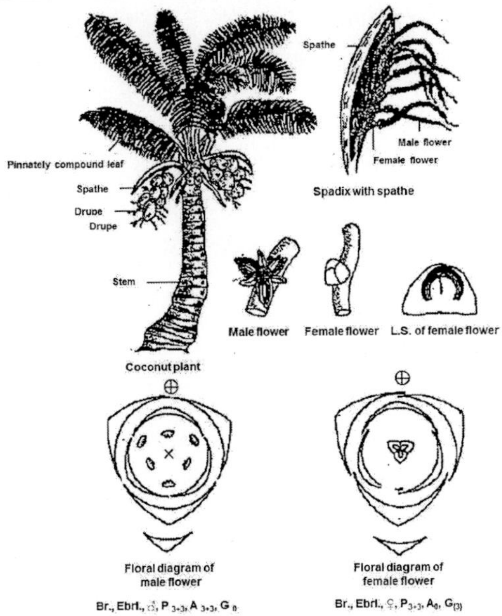

Fig. 12.1. Floral structure and diagram of Cocos nucifera (Arecaceae)

Flower	:	Female flower: Basal portion of spike consists of female flowers
Perianth	:	Tepal - Six arranged in two whorls of equal size, Polyphyllous, imbricate in both whorl, Inner posterior tepal is completely outside
Fruit	:	A fibrous drupe differentiated into exocarp, fibrous mesocarp and hard endocarp enclosing the liquid and solid endosperm
Pollination	:	Cross pollination due to protandrous

Origin: Indo Pacific Region

Related Genera: Alecastrum and Jubaeopsis

Economic Importance

It provides nutritious food and refreshing drink, oil for edible and non edible uses, fibre for commercial value, shell for fuel and industrial uses, leaves for thatch, sap from inflorescence for alcoholic beverage and a variety of miscellaneous products for use as domestic use.

Classification

Coconut is Monotypic: There are three botanical varieties of *Cocos nucifera*,. There are:

(i) Tall: *Cocos nucifera var. typica*

This is a hardy type with the life span of 80 to 90 years. It has long and straight trunk growing to a height of about 15 to 18 meters. The nuts mature within a period of 12 months after pollination. The copra, oil and fibre are of good quality. This variety is cross pollinated because male flowers open first and female flowers are receptive one week after male flowers open.

Eg. West coast call, East coast tall, Kappadam, Andaman Giant, Laccadive Ordinary.

(ii) Semi tall: *Cocos nucifera var. javanica*

This is self pollinated because of overlapping of male and female phases.

(iii) Dwarf type: *Cocos nucifera var. nana*

Characterized by its short stature and earliness in bearing. Life span is less than tall. The endosperm not suitable for oil extraction. Self pollination in nature because of overlapping of male and female phases.

Eg. *Chowghat dwarf, Gangabondm.*

Hybrid palm: The promising hybrid combinations are tall x Dwarf and Dwarf x tall. Hybrids come to flower in four years and perform better in productivity in form of number of nuts per palm and uniformity in bearing.

BOTANICAL FEATURES OF OIL PALM

- Palm oil obtained from freshly fibrous mesocarp (45 to 55%).
- Palm kernel oil is extracted from endosperm

Oilpalm: *Elaeis guineensis* $2n = 32$ **Family:** Arecaceae / Palmae

Habitat	:	Tropical and subtropical
Habit	:	Erect, grows to a height of 20 m at maturity, Monoecious
Root	:	Fibrous root
Stem	:	Trunk is characterized by persistent spirally arranged leaf bases
Leaves	:	Pinnately compound leaves, Petiolate, Exstipulate, Spirally arranged on the top of the trunk, Leaf segment: Sessile, Linear, Entire, Cuspidate, Parallel venation
Inflorescence	:	Compound spadix. Both male and female primordial initiated in each leaf axil, but expressed based on environment. In very drought conditions, male flower get expressed Male Spadix: There is a short stout peduncle supported by mass of finger like spikes each one with as many as 1200 closely packed minute male flowers sunk in tissues of rachis Female spadix: Peduncle is short and stout, carries upto 150 branches each with 20-30 widely separated flowers. Each flower has one bract with acuminate tip; two bracteoles, six perianth, ovary three celled with a short style divided at the top into three stigmatic lobes

Fruit	:	Small, ovoid drupe, borne in large bunches with 200-2000 fruits per bunch, varying in colour. Oil from mesocarp as well as endosperm
Pollination	:	Cross pollination

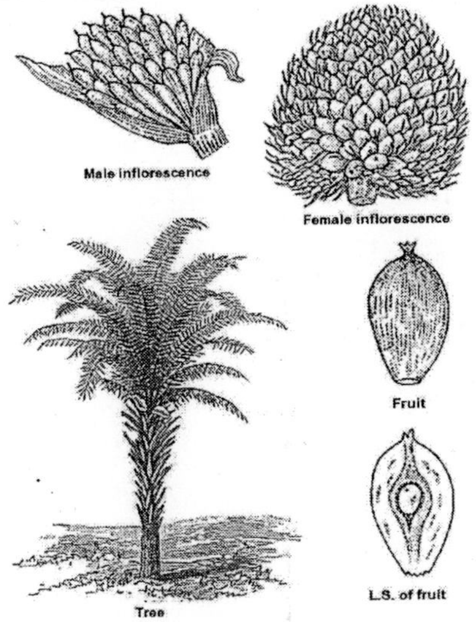

Fig. 12.2. Habit and floral structure of Elaeis guinensis (Areaceae)

Origin: West Africa

Related species: *Elaeis melanococca* and *Elaeis madagascariensis*

Classification:

Elaeis guineensis is monoecious, cross pollinated. Based on shell thickness oil palm cultivars are classified into four varieties.

(i) *Elaeis guineensis var. mesocarpa;*
Fruit consists of 40-60 per cent shell

(ii) *Elaeis guineensis var. dura:*
Shell occupies 20-40 per cent of furit weight

(iii) *Elaeis guneensis var. tenera:*
Shell is thin, occupies 5-20 per cent of furit weight.
High mesocarp content (60-95%)

(iv) *Elaeis guineensis var. Pisifera:*
Shell less (<5%), pea like kernels.

Fruit: A drupe. Both mesocarp and endosperm contains oil. Two distinct oils are extracted.

1. **Palm oil:** Extracted from fleshy mesocarp. The oil content is about 45-55 per cent. The colour is deep orange, containing a carotene precursor for vitamin A. It is more or less solid at ordinary temperature.
2. **Palm kernel oil:** Extracted from kernel, containing oil about 50 per cent. Shells are separated by milling and oil extracted by solvent extraction The oil is liquid at room temperature and is colorless or light yellow in colour. Palm oil is to be extracted immediately after harvest (within 24 hrs) otherwise oil will develop rancidity.

Extraction of oil

(i) **Soft oil process:** Soon after harvest the ripe fruits are boiled in water and pounded to disintegrate the pulp form nut. Again, the pulp is re-cooked with large volume of water. The oil floats on the water and it is skimmed off. The oil thus obtained is good in quality and contains low free fatty acids.

(ii) **Hard oil process:** Here, the disintegration of pulp is by fermentation. Then, pulp is cooked in water and oil is separated. It contain more free fatty acids.

Economic importance

Palm oil and palm kernel oil

- Cooking media
- Soap making, lubricant
- Fuel for internal combustion engines
- Oil cake – cattle feed
- Palm wine by the tapping of male inflorescence

BOTANICAL FEATURES OF ARECANUT

Arecanut (Betelnut): *Areca catechu* $2n = 32$ **Family:** Acecaceae (Palmae)

Habitat	:	Tropical
Habit	:	Slender, unbranched tree
Root	:	Fibrous root
Stem	:	Straight, thin
Leaves	:	Pinnately compound leaf, spirally arranged crowded at the top, leaf sheath broad, thickly coriaceous, embrace the trunlk like a tube, petioles short
Inflorescence	:	Monoecious spadix, axillary, spathes thick coriaceous, after anthesis longitudinally splits and falls off

Flowers	:	Flowers: monoecious, sessile Male flower: perianth six in two whorls, minute, stamens six connate at base, anthers linear lanceolates with a sagitate. Female flowers: triangular perianth six, orbicular, ovate, ovary tricarpellary
Floral formula &diagram		: A in fig 12.3
Fruit	:	A fibrous drupe with ruminate endosperm. It contains alkaloid acrecoline. It is used in veterinary medicine also

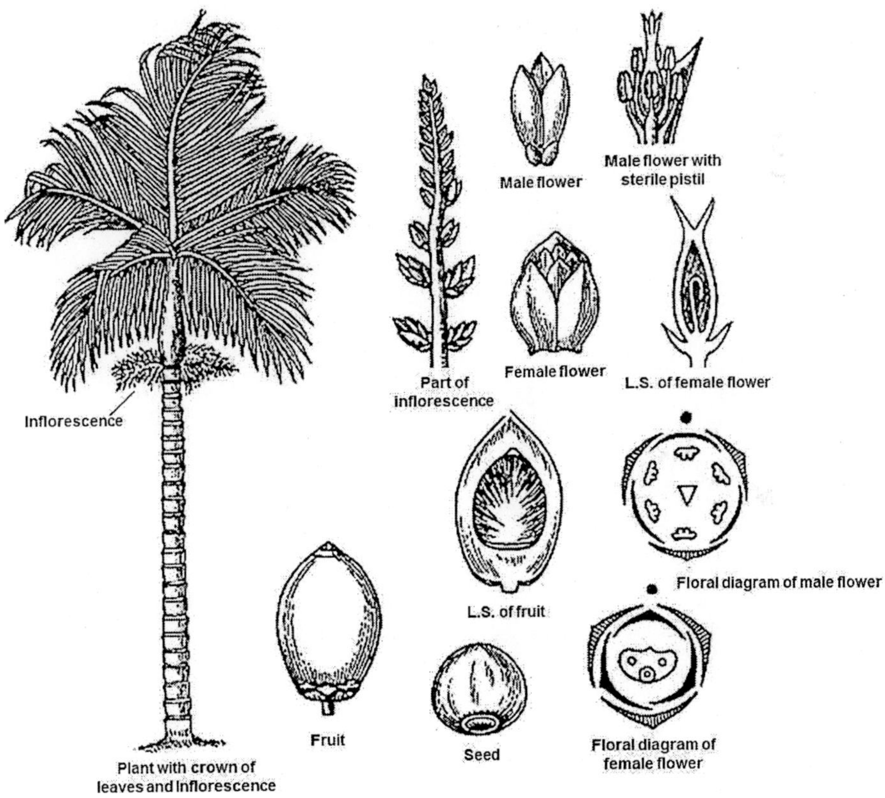

Fig.12.3. Floral structure and diagram of Areca catechu (Arecanut)

Pollination	:	Cross pollination

Economic importance:

- The hard dried endosperm of ripe and unripe seed is chewed as a narcotic.
- It has a number of medicinal uses like an astringent.
- Tree is also grown for ornamental purpose.

1. BOTANICAL FEATURES OF SUGAR PALM

Sugar palm: *Borassus flabellifer* (2n = 36) **Family:** Arecaceae

Habitat	:	Tropical
Habit	:	Tall, stout, grayish black, unbranched
Leaves	:	Simple palmately compound leaf, fan – shaped, petiole spinous
Inflorescence	:	Dioecious, branched, spadices, peduncle sheathed with open spathe Male flowers small, mixed with scaly bracts, sepals three, imbricate, petals three, shorter obovate – spathulate, imbricate, stamens six, pistilode of three bristles Female flowers large, globose , perianth fleshy – sepals, reniform, imbricate, petals three, smaller and convolute, staminode, ovary tri-carpellary, stigma three
Floral formula &diagram	:	As in fig 12.4
Fruit	:	A large subglobose drupe

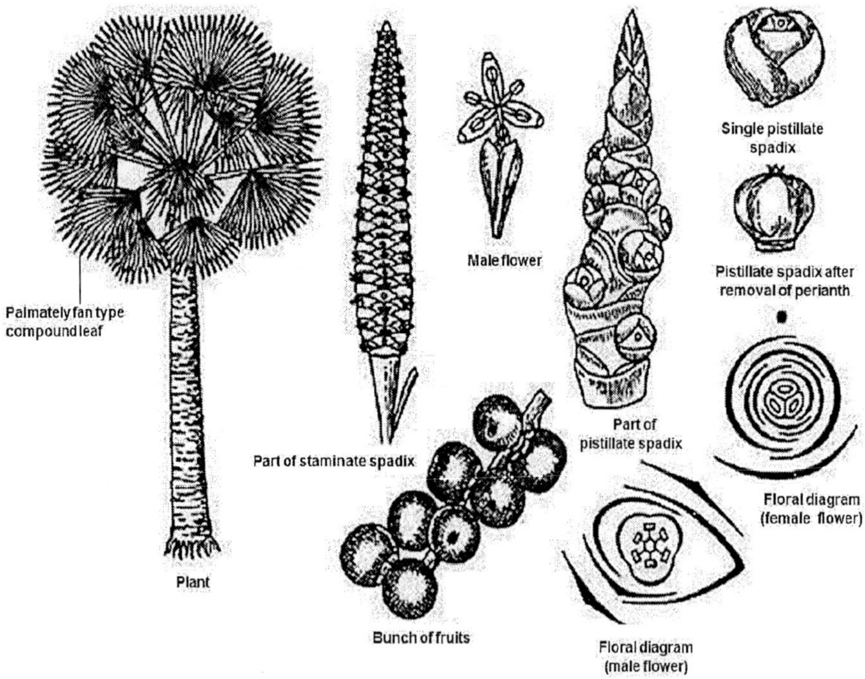

Fig.12.4. Floral structure and diagram of Borassus flabellifer (Arecaceae)

Pollination : Cross pollination
Place of origin : Africa
Distribution : India, Burma, Srilanka, Malaysia

Extraction of sugar:

Neera tapped from unopened inflorescence. Generally the top is cut off and the sap oozes out and is collected in containers. The yield of neera amounts to 3 or 4 qt. a day for several months. The sap possesses sugar content of about 14 per cent. The sweet juice is boiled down to a syrupy consistency and pressed into leaves to cool and harden into the crude sugar, known as jaggery.

Economic importance:

Juice from male peduncle used as toddy; fermented into arrack, treated with lime and boiled down into jiggery. Wood is used as rafters, posts, fences. Leaves are used for thatching fans, hats, baskets, etc.

- Neera contains 14 per cent sucrose.
- Toddy – fermented neera.
- Palm jaggery – by boiling neera
- Leaf – for brushes and brooms
- Leaf – Baskets, mat etc.,
- Timber – rafters, pillar, posts, fuel.

LIST OF CULTIVATED CROPS UNDER FAMILY MALVACEAE

Sl. No.	List of cultivated crops	Scientific name	Chromosome number	Economic Parts
	Fiber Crops			
1	Cotton	*Gossypium herbaceum*	26	Lint
		Gossypium arboreum	26	Lint
		Gossypium hirsutum	52	Lint
		Gossypium barbadense	52	Lint
2	Mesta	*Hibiscus cannabinus*	36	Stem
	Vegetables			
1	Bhendi	*Abelmoschus esculentus*	72-132	Fruit

Family description of Malvaceae

Plant	:	Annual or perennial herb / Shrub / Tree
Stem	:	Erect, cylindrical, Solid, Branched
Leaf	:	Simple, Petiolate, Stipulate, alternate
Inflorescence	:	Axillary or terminal solitary Cyme
Flower	:	Pedicellate, Actinomorphic (except Bhendy), Bisexual, epicalyx present, Hypogynous
Calyx	:	Sepals – 5, gamosepalous, valvate aestivation
Corolla	:	Petals – 5, Polypetalous, Twisted aestivation
Androecium	:	Stamens – numerous, Monadelphous, Reniform anthers
Gynoecium	:	Carpels – 5, Syncarpous, Superior ovary, Penta locular with many ovules in each locule with axile placentation
Fruit	:	Capsule

Systematic position of family Malvaceae

Kingdom	:	Plant kingdom
Division	:	Phanerogams
Subdivision	:	Angiosperms
Class	:	Dicotyledonae
Subclass	:	Polypetalae
Series	:	Thalamiflorae
Order	:	Malvales
Family	:	Malvaceae

Classification of Fibre based on botanical origin

Bast Fibre (Stem / soft phloem)	**Structural Fibre (hard / leaf)**	**Surface Fibre (Seed / fruit)**
• Derived mostly from stems of dicotyledonous plants.	• Derived from leaves of the monocotyledonous plants.	• Fibres are arising as single celled outgrowths from the seeds or inner walls of the fruits
• Fibres are usually separated by 'retting'.	• These fibres are highly lignified and separated by 'scrapping'	• Fibres are usually separated by 'ginning'
• Eg. Jute, hemp, kenaf, roselle and ramie	• Eg. Manila hemp, sisal, sansevieria	• Eg. Cotton, Kapok.

BOTANICAL FEATURES OF COTTON

Cotton: *Gossypium spp.* $2n = 26, 52$ **Family:** Malvaceae

Habitat	:	Tropical and subtropical
Habit	:	Herbaceous annual growing to a height of 1-1.5 m primary axis is erect and branched.
Root	:	Tap root
Stem	:	Branches are dimorphic in nature. 1. Vegetative axillary bud (monopodial,) 2. Fruiting extra axillary bud (sympodial) in upper nodes. Vegetative branches are monopodial and they are morphologically similar to main stem. They do not bear flowers directly, but gives out secondary branches. Fruiting branches are sympodial. They arise from the axes of subtending leaf, another branch develop terminating in a fruiting point
Leaves	:	Simple, Petiolate, Stipulate, Alternate, Cordate with 3-9 lobed and palmately veined. The size, texture, shape and hairiness are widely varying
Inflorescence	:	Simple solitary Cyme
Flower	:	Pedicellate, Bracteoles called epicalyx, three in number deeply toothed persistant at the base of the flower. Actinomorphic, Bisexual, Complete, Pentamerous, Hypogynous
Calyx	:	Sepals – 5, Gamosepalous, cup shaped, Valvate
Corolla	:	Petals – 5, polypetalous, Twisted aestivation, Petals united to the staminal tube at the base (Epipetalous). A purple spot is found on the inner side of the claw of the petal (petal spot) in some species viz., *G. barbadense*
Androecium	:	Stamens unit to form staminal column bearing many filaments ending with reniform anthers, Polyandrous, Monadelphous, Epipetalous.
Gynoecium	:	Pentacarpellary, Syncarpous, Superior ovary, Many ovules in axile placentation, style slender passes through staminal column with united lobed stigma.
Fruit	:	Capsule called boll. Seeds 1-9 in each locule, subglobose seeds are covered with one to two layers of hairs. Lint arises from the epidermal layer and fuzz arises from the sub epidermal layer. Fuzz cannot be separated from the seed. Lint is separable
Floral Formula &	:	As in fig 12.5

Floral Diagram :

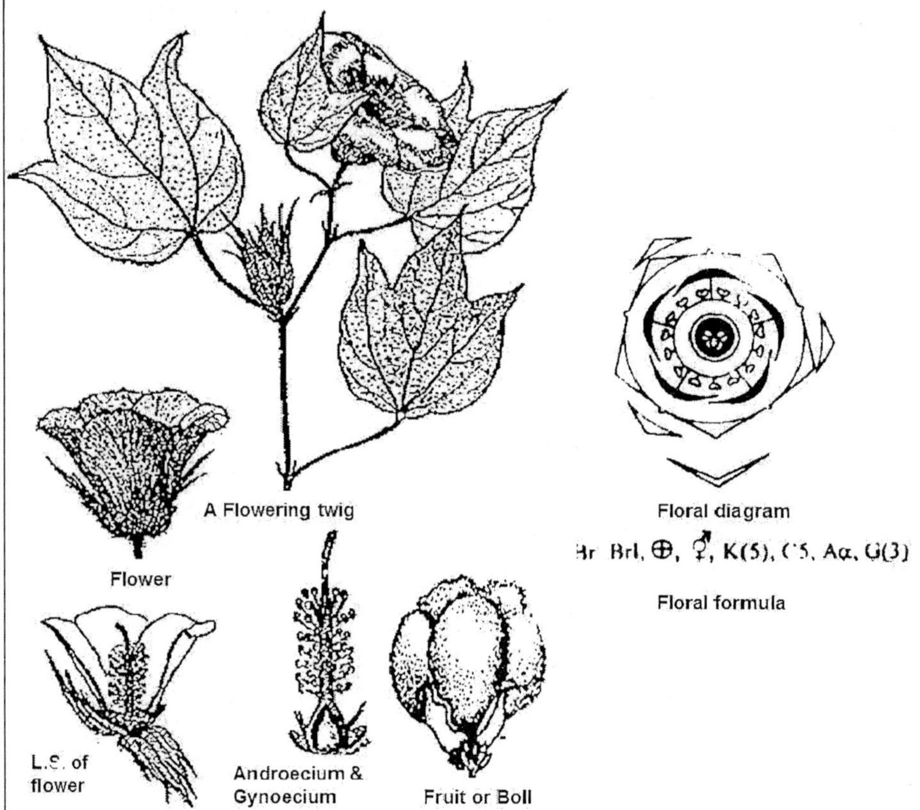

Fig. 12.5. Floral structure and diagram of Gossipium barbadense (Malvaceae)

Pollination	:	Often cross pollination

Origin:

Gossypium herbaceum : Africa and Asia
Gossypium arboreum : Asia
Gossypium hirsutum : Mexico
Gossypium barbadense : West Indies

Economic Importance:

- Cotton lint is a valuable textile. Lint is used in textile, absorbent cotton, wiping, polishing material, clothing, house hold articles, industrial articles, twines, cellulose and plastic.
- In industries it is used for making tyre cord and machinery belt.
- Cotton seed is used as cattle feed, cotton seed oil is used as oil which has 20 per cent semi drying oil.
- Stalk is used as fuel, paper pulp, bast fibre extracted. Leaves are used as fodder.

Four different cultivated species of cotton

1	Character	G.herbaceum	G. arboretum	G.hirsutum	G. barbadense
1.	Genome	A_1	A_2	$(AD)_1$	$(AD)_2$
2.	Races	*africanum* *acerifolium* *persicum* *kuljianum* *weightianum*	*indicum* *burmanium* *cernnum* *sinense* *bengalen* *sesudenense*	*punctatum* *latifolium* *marie-galante*	*brasiliense* *darwinii*
3.	Origin	Africa & Asia	Asia	Mexico	West Indies
4.	Leaves	Cordate, 3-5 deeply lobed	Palmately, deeply lobed, green	3-5 lobed, large, very dark green	3-5 lobed, large, cordate, lobes triangular
5.	Bracteoles	Entire or denate, 3-4	Entire or denate, 3-4	Deeply lanceolate, 8-12 tooth	Deeply lanceolate, 8-12 tooth
6.	Nectar	Absent	Absent	Present	Present
7.	Petals	Pale yellow with reddish spot	Pale yellow to cream	Bright yellow	Pale yellow
8.	Capsule	Round, beaked with few oil glands	Tappering, pitted with oil glands	Medium round with few oil glands	Long, broad at base, tip acute, oil glands rough
9.	Seeds	With short fuzz	With fuzz and lint	Even coat of lint and full coat of fuzz	Moderate lint and thick coat of fuzz
10.	Lint	Short stapple	Short stapple	Long stapple	Medium stapple

BOTANICAL FEATURES OF MESTA (*Kenaf*)

The bast fibre obtained from the stem is used in a similar manner to jute

Mesta (Kenaf): *Hibiscus cannabinus* (2*n* = 36) **Family:** Malvaceae

Habitat	:	Tropical / Subtropical
Habit	:	Herbaceous annual growing to a height of even 5 metres
Root	:	Tap root

Stem	:	Straight with small prickles
Leaves	:	Simple, Petiolate, Alternate, stipulate (stipules long and pointed), Lanceolate / Lobed, Serrated margins, Acute apex, Reticulate venation
Inflorescence	:	Axillary solitary Cyme
Flower	:	Pedicellate, Bracteoles represented as epicalyx consisting of 8 to 10 bracteoles, Actinomorphic, Bisexual, Pentamerous, Hypogynous
Calyx	:	Spals – 5, Gamosepalous, Valvate, Persistent
Corolla	:	Petals – 5, Polypetalous, Twisted aestivation, Pale yellow to sulphur colour with crimson or purpled center
Androecium	:	Stamens numerous on a staminal column, Polyandrous, Monadelphous, Epipetalous, Reniformed anther
Gynoecium	:	Penta-carpellary, Syncarpous, Superior ovary, style passes through staminal column and terminates in five lobed stigma
Fruit	:	Capsule with persistent calyx, five loculed, each locule containing four to five ovules in axile placentation. Capsule pointed and bristled, seed glabrous
Floral Formula &	:	As in fig.12.6

Floral Diagram :

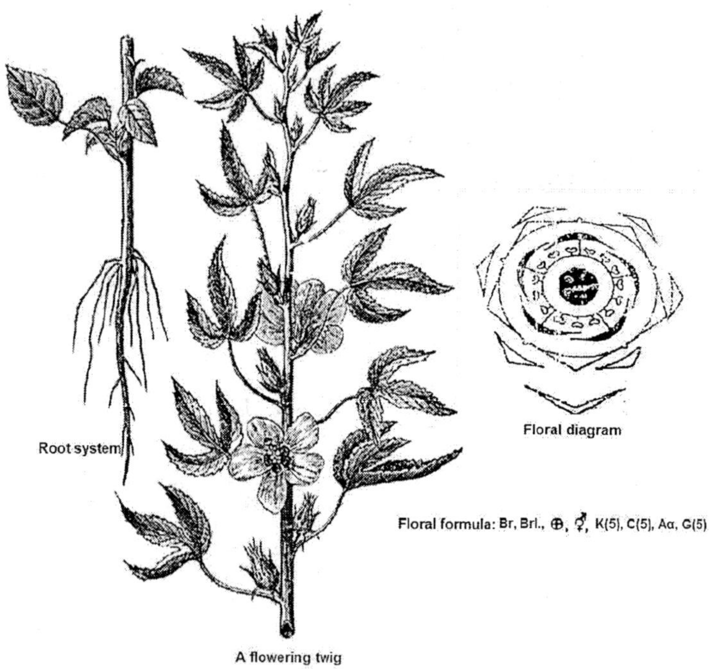

Fig. 12.6. Floral diagram and formula of Hibiscus cannabinus (Malvaceae)

Origin: Africa

Economic Importance:

- Fibre is mixed with jute for manufacture of bags, sacks, netting, opes.
- It is used in the manufacture of sand paper and abrasives.
- The seeds contain about 20 per cent oil which is sometimes extracted and used as a lubricant and for illumination; for manufacture of soap, linoleum and in paints and varnishes

BOTANICAL FEATURES OF BHENDY (*Okra*)

It is important vegetable crop grown in tropical and subtropical parts of the world

Bhendy (Okra): *Abelmoschus esculentus* ($2n = 72 - 132$) **Family:** Malvaceae

Habitat	:	It is cultivated throughout the tropical and warm temperate regions of the world
Habit	:	Annual herb
Root	:	Well developed deep tap root
Stem	:	Erect, woody, Branched
Leaves	:	Simple, Petiolate, Stipulate, Spirally arranged, leaves are broad, cardatewith with 5–7 lobes, Palmately reticulate, Pubescent
Inflorescence	:	Axillary solitary Cyme
Flower	:	Pedicellate, Epicalys consists of a whorl of 10 hairy bracteoles being persistent, Zygomorphic with regard to arrangement of sepals, Bisexual, Pentamerous, Hypogynous
Calyx	:	Sepals – 5, Gamosepalous, Valvate aestivation
Corolla	:	Petals - 5, Polypetalous, Twisted aestivation Petals united with staminal tube at the base
Androecium	:	Stamen - numerous, monadelphous, filaments united in a tube and basally joined to corolla (Epipetalous), anthers unilocular, reniform
Gynoecium	:	5 or 6 carpels, Syncarpous, superior ovary, 5 to 6 locules with one or two ovules in each locule in axile placentation Stigma passes through staminal column
Fruit	:	Capsule
Floral Formula &	:	As in Fig 12.7

Floral Diagram

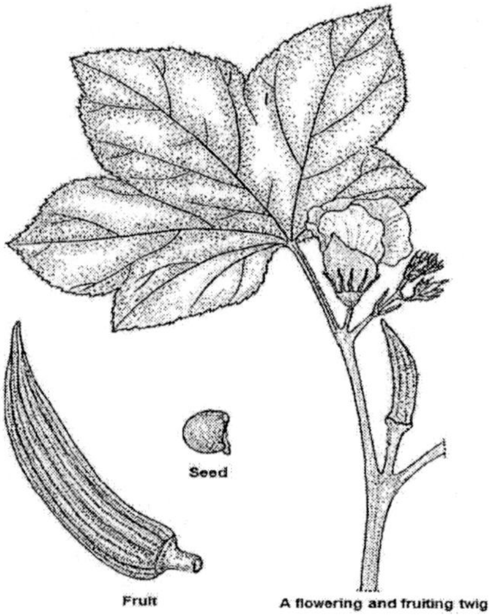

Fig. 12.7. Bhendi: Abelmoschus esculentus (Malvaceae)

Pollination	:	Self pollination
Origin	:	South East Asia

Economic importance

- Green pods are used as vegetable.
- The immature pod can be used in pickle making.

QUESTION BANK: UNIT- IV

BOTANICAL DESCRIPTION AND ECONOMIC USES OF PEDALIACEAE, ASTERACEAE, OLEACEAE, BRASSICACEAE, EUPHORBIACEAE, ARECACEAE AND MALVACEAE

PART-A

I. CHOOSE THE CORRECT ANSWER 0.5 MARKS

					Key
1.	Biodiesel yielding plant				
	a) Groundnut	b) Jatropha	c) Castor	d) Gingelly	b
2.	Coconut toddy is derived from				
	a) Coconut water	b) Inflorescence	c) Coconut milk	d) None of them	b
3.	The economic part of jasmine is				
	a) Seed	b) Leaf	c) Flower	d) Fruit	c
4.	Jasmine belongs to the family of				
	a) Malvaceae	b) Brassicaceae	c) Poaceae	d) Oleaceae	d
5.	Surface or seed fibre obtained from				
	a) Cotton	b) Mesta	c) Jute	d) All of them	a
6.	Stem or Bast fibre obtained from				
	a) Jute	b) Mesta	c) Ramie	d) All of them	d
7.	Structural or leaf fibre obtained from				
	a) Manila hemp	b) Sun hemp	c) Mesta	d) Cotton	a
8.	All stamens fused form a tube as in cotton is called				
	a) Polyadelphous	b) Monoadelphous	c) Diadelphous	d) None of them	b
9.	Which of the following district having greater number of sago factories in Tamil Nadu				
	a) Ariyalur	b) Thanjavur	c) Salem	d) Vellore	c
10.	Inflorescence of sunflower is				
	a) Cyanthiyum	b) Capitulum	c) Head	d) Both b & c	d
11.	The fruit type of sunflower is				
	a) Caryopsis	b) Regma	c) Utricle	d) Achene	d
12.	Chromosome number of sunflower is				
	a) 2n=26	b) 2n=24	c) 2n=34	d) 2n=40	b
13.	Which of the following is putative parent for cultivated sunflower				
	a) *H. hirsutus*	b) *H. rigidus*	c) *H. petriolaris*	d) All of them	c
14.	The centre of origin of sunflower is				
	a) Africa	b) India	c) North America	d) South America	c

15. Chromosome number of gingelly or sesamum is

a) 2n=26	b) 2n=30	c) 2n=33	d) 2n=40	**a**

16. The androecium of gingelly is

a) 4 stamens	b) Didynamous	c) Epipetalous	d) All of them	**d**

17 Secondary centre of orgin of gingelly is

a) India	b) China	c) Africa	d) Sudan	**a**

18. The primary centre of origin of gingelly is

a) Africa	b) India	c) China	d) Europe	**a**

19. The oil content of gingelly seed is

a) 25-35%	b) 35-45%	c) 45-55%	d) 55-65%	**c**

20. The androecium of mustard is

a) 6 stamens	b) Tetradynamous	c) Polyandrous	d) All of them	**d**

21. Fruit type of mustards is

a) Siliqua	b) Regma	c) Utricle	d) Achene	**a**

22. The botanical name of Indian mustards is

a) *B. campestris*	b) *B. nigra*	c) *B. alba*	d) *B. juncea*	**d**

23. Fruit type of castor is

a) Siliqua	b) Regma	c) Utricle	d) Achene	**b**

24. The chromosome number of coconut, oil palm and arecanut are

a) 2n=36	b) 2n=30	c) 2n=32	d) 2n=40	**c**

25. Origin of *Gossypium hirsutum* is

a) Mexico	b) China	c) West indies	d) Africa	**a**

26. Friut type of mesta is

a) Siliqua	b) Capsule	c) Utricle	d) Achene	**b**

II. FILL IN THE BLANKS

1. Tapioca is rich in ---------------------
2. ------------------is polyamide nylon manufactured from castor oil fraction
3. Arecanut contains condensed tannins are called ---------------------
4. Fruit type of Euphorbiaceae family is----------------------
5. Pyrethrum is a natural insecticide derived from------------------
6. Botanical name of upland cotton is-----------------------
7. Botanical name of sunflower is------------------
8. Cotton fibre is alomost pure-----------------------
9. Stem or Bast fibre obtained from---------------, ------------- and ------------
10. Structural or leaf fibre obtained from------------ and ----------------
11. Surface or seed fibre obtained from ---------------- and -------------
12. β-carotene and β-cryptoxanthin is precursor for ----------------

13. Botanical name of Arecanut is -------------------
14. Inflorescence of sunflower is known as ----------------------
15. Tapioca processing industry is also called as----------------
16. ----------------- is the bitter glycoside presented in Corchorus capsularis.
17. The 2n number of gingelly is ------------
18. Fruit type of gingelly is -----------------
19. The secondary centre of origin of gingelly is --------------
20. Botanical name of safflower is ------------------------
21. The oil content of sunflower is -----------------
22. The natural insecticide pyrethrum is derived from ------------------------
23. The economic part of cabbage is -------------------------
24. Botanical name of cabbage is ---------------------------
25. Botanical name of cauliflower is --------------------------
26. Economic part of coconut is ---------------------------
27. Economic part of oil palm is ------------------ and --------------
28. Botanical name of oil palm is ---------------------
29. Botanical name o arecanut is --------------------
30. Inflorescence of coconut is -----------------------

Answer keys:

1. Starch	2. Rilson	3. Arecatannins
4. Regma	5. *C. cinerarifolium*	6. *G. hirsutum*
7. H. annus	8. Cellulose	9. Jute, Mesta and Ramie
10. Manila hemp and Sisal		11. Cotton and Kapak
12. Vitamin A	13. Areca catechu	14. Head or capitulum
15. Sago factories	16. Corchorin	17. 26
18. Loculicidal capsule	19. India	20. *Carthamus tinctorius*
21. 25-35%	22. *Chrysanthemum cinerariaefolium*	
23. Leafy head	24. *Brassica oleraceae var. capitata.*	
25. *Brassica oleraceae var. botrytis.*		
26. Endosperm		
27. Kernel and Mesocarp		
29. Elaeis guinensis	30. Areca catechu.	31. Compound spike.

III. MATCH THE FOLLOWING

1.	Sunflower	(a) Seed fibre
2.	Tapioca	(b) Stem fibre
3.	Cotton	(c) Regma
4.	Jute	(d) Sago factories
5.	Castor	(e) Capitulum

Answer key:

1. (e) 2. (d) 3. (a) 4. (b) 5. (c)

IV. STATE TRUE OR FALSE

1. Sunhemp is example for green manure and stem fibre.
2. Jute belongs to the family Tilliaceae.
3. Inflorescence of Sunflower is panicle.
4. The calyx and corolla of pedaliaceae family is gamosepalous and gamopetalous
5. Sesamum is often cross pollinated crop.
6. The 2n number of sunflower is 36.
7. India is the secondary centre of origin for gingelly
8. The fruit type of sunflower is achene.
9. The natural insecticide pyrethrum is derived from Chrysanthemum indicum.
10. The economic part of cauliflower is curd.
11. Castor is a protogynous crop.
12. The botanical name of palmyra palm is Borassus flabellifer.
13. The origin of coconut is North America.
14. Bast fibre extracted from stem of Mesta.

Answer key:

1. True 2. True 3. False 4. True 5. True 6. True
7. True 8. True 9. False 10. True 11. True 12. True
13. False 14. True.

PART-B

1 MARK

1. Origin of *G. hirsutum*
2. What is ginning percentage?
3. Uses of wild species for cotton improvement.
4. Classification of commercial cotton based on fibre length.
5. Inflorescence of coconut.

6. List out the cotton wild species.
7. Uses of cauliflower.

PART-C

2 MARK

1. Economic importance of gingelly.
2. Write short notes on structure of tapioca (cassava) tuber.
3. Extraction of sugar from palmyra palm.
4. Origin of tetraploid cotton (G. hirsutum).
5. List out two fibre crops, their scientific name, family and economic parts.
6. Write about wild speices utilized for cotton improvement.
7. Illustrate the genetic relationship (U triangle) of oilseed Brassica by Nagaharu U (1935).
8. Distinguishing characters of Pedaliaceae family.
9. Distinguishing characters of Arecaceae family.
10. Classification of cultivated coconut.
11. Write about inflorescence of castor.
12. Write uses of oil palm.
13. Elucidate the economic importance of sunflower.
14. Write uses of castor.
15. Write utilization of palmyra palm.

PART-D

5 MARK

1. What is tapping for coconut toddy? Write economic importance of coconut.
2. Write elaborately economic importance of palmyra palm.
3. Describe botanical features of coconut.
4. Describe botanical features of sunflower.
5. Explain characteristics of pedaliaceae family
6. Elucidate the economic importance of oil palm
7. Write economic importance of cotton and jute.

Chapter

13

List of cultivated Crops, Economic parts, Chromosome Number and Family description of Tiliaceae, Piperaceae and Chenopodiaceae

LIST OF CULTIVATED CROPS UNDER FAMILY TILIACEAE

Sl. No.	List of cultivated crops	Scientific name	Chromosome number	Economic Parts
	Fiber Crops			
1	White Jute	*Corchorus capsularis*	14	Stem
2	Tossa Jute	*Corchorus olitorius*	14	Stem

Family description of Tiliaceae

Plant	:	Annual herb / under shrub
Stem	:	Erect, Tall
Leaf	:	Simple, Petiolate, alternate, stipulate
Inflorescence	:	Axillary solitary Cyme
Flower	:	Pedicellate, Bracteate, Actinomorphic, Bisexual, Hypogynous
Calyx	:	Sepals – 5, Polysepalous, Valvate
Corolla	:	Petals – 5, Polypetalous, Valvate
Androecium	:	Stamens – many, Polyandrous, Anthers – two celled
Gynoecium	:	Carpels – 2 to 10, Syncarpous, Superior ovary
Fruit	:	Capsule

Systematic position of family Tiliaceae

Kingdom : Plant kingdom

Division : Phanerogams

Subdivision : Angiosperms

Class : Dicotyledonae

Subclass : Polypetalae

Series	:	Thalamiflorae
Order	:	Malvales
Family	:	Tiliaceae

BOTANICAL FEATURES OF JUTE

- Jute is extensively used for the manufacture of gunny bags as packing materials.
- Jute fibre is obtained from the bark of the plant which is the secondary phloem fibres of the stem.

White Jute: *Corchorus capsularis*

Tossa jute: *Corchorus olitorius* $2n = 14$ **Family:** Tiliaceae

Habitat	:	Deltaic area at low altitude with high temperature and high humidity
Habit	:	Erect annual herb / Under shrub
Root	:	Tap root
Stem	:	Erect, Cylindrical, Solid, Branched
Leaves	:	Simple, Petiolate, Alternate, Stipulate, Lanceolate leaf shape, Serrated margin, Acuminate apex, Reticulate venation
Inflorescence	:	Axillary solitary Cyme
Flower	:	Pedicellate, Bracteate, Actinomorphic, Bisexual, Hypogynous
Calyx	:	Sepals – 5, Polysepalous, Valvate aestivation
Corolla	:	Petals – 5, Polypetalous, Valvate aestivation
Androecium	:	Stamens – 15 to 20, Five in each whorl, Polyandrous, Dorsifixed
Gynoecium	:	Pentacarpellary, Syncarpous, Superior ovary, Pentalocular, many ovule sin axile placentation, Short style, flattened stigma
Fruit	:	Capsule dehiscing longitudinally
Pollination	:	Self pollination
Floral Formula	:	As in fig.13.1
Floral Diagram	:	As in fig.13.1

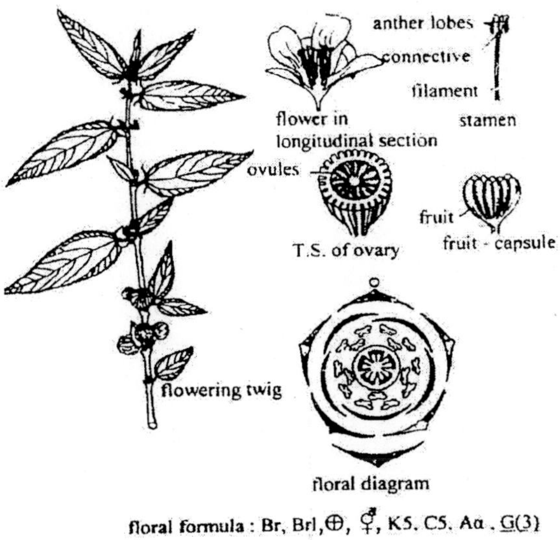

Fig. 13.1. Floral structure and diagramof Corchorus olitorius (Tilliaceae)

Origin:

White Jute: *Corchorus capsularis* – South Asia

Tossa jute: *Corchorus olitorius* – Africa and Asia

Cultivated Species of Jute

S.N.	Characters	*Corchorus capsularis*	*Corchorus olitorius*
1	Habitat	Low land	High land
2	Origin	South Asia	Africa
3	Variability	Wide	Narrow
4	Leaves	Bitter taste	Sweet taste
5	Direction of leaf spiral	Right	Left
6	Flower	Smaller	Larger
7	Stamens	20-30	30-60
8	Capsule	Globular	Cylindrical
9	Fiber colour	White	Golden brown
10	Fiber quality	Low	High

Economic Importance

- Jute is used chiefly for rough weaving. The thick cloth made from jute is used for making gunny – bags, sails for country boars. Another type of fine cloth prepared from jute fibre is chiefly used as a cloth to sleep on.

- Jute is also extensively used in manufacture of carpets, curtains, shirtings and twine and ropes.
- Jute butts (short fibres) are used for making paper.

Retting of Jute (Extraction of jute fibre)

Retting is a process by which the fibre in the bark get loosened and separated from the woody stack due to the removal of pectin's, gums and other mucilaginous substance. This is usually effected by the combined action of water and microorganisms. The tied bundles of jute stems are taken to the nearest pool or ditch for retting. The bundles are laid flat in water at least 0.6 to 0.9 m in depth and arranged side by side so as to form a regular platform, which is usually known as jak. The retting process is completed in 8-30 days of fibre is extracted from the stalks of retted jute by hand or by decorticators. Retting can also be done by chemical process.

List of cultivated crops under family Piperaceae

Sl. No.	List of cultivated crops	Scientific name	Chromosome number	Economic Parts
	Mastigatory			
1	Betel vine	Piper betle	32	Leaves

Family description of Piperaceae

Plant	:	Creeper
Stem	:	Swollen nodes
Leaf	:	Alternate, Fleshy, Cardate
Inflorescence	:	Fleshy spike
Flower	:	Bisexual / Unisexual
Perianth	:	Absent
Androecium	:	Stamens – 1 to 10
Gynoecium	:	One celled ovary with a basal ovule
Fruit	:	A drupe

Systematic position of family Piperaceae

Kingdom	:	Plant kingdom
Division	:	Phanerogams
Subdivision	:	Angiosperms
Class	:	Dicotyledonae
Series	:	Magnoliids
Order	:	Piperales
Family	:	Piperaceae

BOTANICAL FEATURES OF BETEL VINE

Betel vine is grown under shade with protection from wind. The leaves are chewed together with betel nut as a mastigatory

Betel vine: *Piper betle* $2n = 32$ **Family:** Piperaceae

Habitat	:	Tropical
Habit	:	A perennial creeper, Branches with swollen nodes
Root	:	Tap root
Stem	:	Slender, twining
Leaves	:	Simple, Petiolate, alternate, Stipulate, Fleshy, Ovate / Cardate leaf shape, Entire margin
Inflorescence	:	Fleshy spike (A catkin)
Flower	:	Very minute, Unisexual
Perianth	:	Absent
Androecium	:	Stamen – 1 to 10
Gynoecium	:	Monocarpellary, Superior ovary, Single ovule in basal placentation
Fruit	:	One seeded drupe

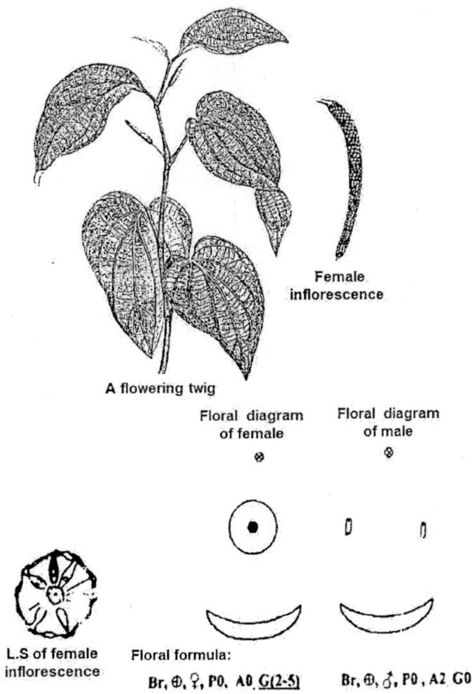

Fig. 13.2. Floral structure and diagram of Piper betel (Piperaceae)

Origin: Malaysia

Related species of betel vine

Indian pepper : Piper longum
Javanese pepper : Piper retrofractum

Uses

- Fresh leaves of betel leaves are smeared with lime and chewed along with areca nut usually after meals.
- Betel leaves are used as medicine in folk medicines

LIST OF CULTIVATED CROPS UNDER FAMILY CHENOPODIACEAE

Sl. No.	List of cultivated crops	Scientific name	Chromosome number	Economic Parts
	Sugar crops			
1	Sugar beet	Beta vulgaris	18	Root tuber

Family description of Chenopodiaceae

Plant	:	Biennial herb
Stem	:	Shortened erect stem, Herbaceous, Cylindrical, Solid
Leaf	:	Fleshy leaves arise in cluster from a shortened stem
Inflorescence	:	Three to four flowered cyme arranged on a spike
Flower	:	Small, Bisexual, Monochlamydeous
Perianth	:	Five, imbricate
Androecium	:	Five anthers
Gynoecium	:	Bicarpellary, Syncarpous, Curved embryo
Fruit	:	Utricle

Systematic position of family Chenopodiaceae

Kingdom	:	Plant kingdom
Division	:	Phanerogams
Subdivision	:	Angiosperms
Class	:	Dicotyledonae
Subclass	:	Monochlamydeae
Series	:	Curvembryeae
Family	:	Chenopodiaceae

BOTANICAL FEATURES OF SUGAR BEET

Suger beet is a white rooted biennial cultivar for sugar production. About one third of world's sugar production is from sugar beet. It develops a large succulent root in the first year and a seed stock in the second year. Occasionally, some plants produce a seed stock in the first year itself called bolting.

Sugar beet: *Beta vulgaris* $2n = 18$ **Family:** Chenopodiaceae

Habitat	:	Tropical / Subtropical
Habit	:	Biennial glabrous herb
Root	:	Tap root
Stem	:	Shortened stem
Leaves	:	Fleshy leaves arise in cluster from a shortened stem
Inflorescence	:	Three to four flowered cyme arranged on a spike; each subtended by a small narrow leaf
Flower	:	Small, Bisexual, Monochlamydeous
Perianth	:	Tepals - Five, imbricate
Androecium	:	Five anthers
Gynoecium	:	Bicarpellary, Syncarpous, Inferior ovary, Curved embryo with three stigma
Fruit	:	Utricle

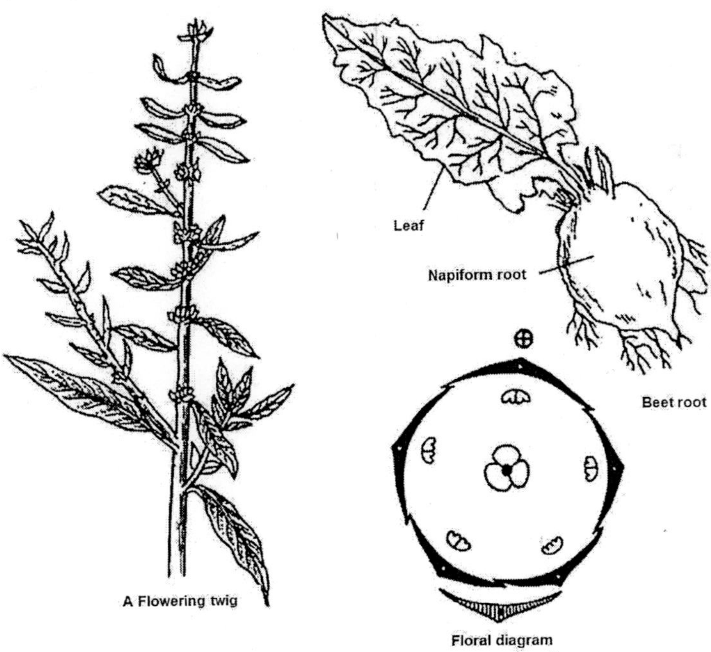

Floral formula: Br, ⊕, ⚥, P_{3-5}, A_5, $G_{(3)}$

Fig. 13.3. Floral structure and diagram of Beta vulgaris (Chenopodiaceae)

Pollination: Cross pollination

Origin: Eurasia

Cultivated types of sugar beet

Beta vulgaris sub sp. vulgaris includes:

1. **Sugar beet:** White rooted; used for sugar production.
2. **Beet root:** Deep red colour, globular or cylindrical in shap; used for vegetable purpose.
3. **Spinach beet:** Leaves have thickened mid ribs; used as spinach.
4. **Chard beet:** Used as cattle feed in Europe.

Extraction of sugar

The sugar is stored in the form of sucrose in the enlarged taproot and hypocotyls (12-20%). The roots are cut into small pieces, treated repeatedly with hot water and sugar is extracted by diffusion process in a series of tanks. The extracted juice is treated with lime to coagulate some of non-sugars and then with carbon dioxide, which precipitates calcium carbonates. The juice is filtered and the resulting liquid is concentrated, crystallized and centrifuged. The beet sugar is identical in composition and appearance to the usual sugar obtained from sugarcane.

Uses of sugar beet

1. Beet sugar is used as a source of sugar.
2. Leaf is used as cattle feed.
3. Filter cake is used as manure.
4. Molasses is used as stock feed and in preparation of ethanol.
5. Spinach beet are used for spinach.
6. Chard types of beet are used as cattle feed.
7. Beet root types are used as vegetables either eaten boiling or cooked.

Chapter

14

List of cultivated Crops, Economic parts, Chromosome Number and Family description of Solanaceae, Mimosae and Moraceae

LIST OF CULTIVATED CROPS UNDER FAMILY SOLANACEAE

Sl. No.	List of cultivated crops	Scientific name	Chromosome number	Economic Parts
	Narcotic and Fumitory			
1	Tobacco	*Nicotiana tabacum*	48	Leaves
	Starch crop			
1	Potato	*Solanum tuberosum*	48	Stem Tuber
	Spices and Condiments			
1	Chillies	*Capsicum annuum*	24	Fruit
	Vegetable Crops			
1	Tomato	*Solanum lycopersicum*	24	Fruit
2	Brinjal	*Solanum melongena (Syn. S. esculentum)*	24	Fruit

FAMILY DESCRIPTION OF SOLANACEAE

Plant	:	Herb / Shrub with watery juice
Stem	:	Erect, Cylindrical, Branched, Herbaceous
Leaf	:	Simple, Alternate, Exstipulate, reticulate
Inflorescence	:	Axillary Solitary Cyme
Flower	:	Pedicellate, Bracteate, Complete, Actinomorphic, Bisexual, Hypogynous
Calyx	:	Sepals -5, Gamosepalous, Persistent, Valvate aestivation
Corolla	:	Petals-5, Gamopetalous, Funnel shaped, Valvate aestivation
Androecium	:	Stamens-5, Epipetalous
Gynoecium	:	Bicarpellary, Syncarpous, Superior ovary, Ovules – numerous in axile placentation
Fruit	:	Berry / Capsule

Systematic position of family Solanaceae

Kingdom	:	Plant kingdom
Division	:	Phanerogams
Subdivision	:	Angiosperms
Class	:	Dicotyledonae
Subclass	:	Gamopetalae
Series	:	Bicarpellatae
Order	:	Polymoniales
Family	:	Solanaceae

BOTANICAL FEATURES OF TOBACCO

In India tobacco is cultivated in almost all the states. Andra Pradesh has the largest area under tobacco followed by Gujarat.

Tobacco: *Nicotiana tobacum* $2n = 48$ **Family:** Solanaceae

Habitat	:	Tropical
Habit	:	Stout, annual herb about three metres high
Root	:	Tap root
Stem	:	Thick, Erect, Unbranched, Basal portion is woody
Leaves	:	Simple, Petiolate, Spirally arranged, Lanceolate, Entire, Acute apex, Reticulate venation
Inflorescence	:	Terminal panicle
Flower	:	Pedicellate, Bracteate, Actinomorphic, Bisexual, Hypogynous
Calyx	:	Sepals – 5, Gamosepalous, Valvate
Corolla	:	Petals – 5, Gamopetalous, Twisted in bud, Valvate in opened flower
Androecium	:	Stamens – 5, Epipetalous
Gynoecium	:	Bicarpellary, Syncarpous, Superior ovary, Ovules – numerous in axile placentation, Capitate stigma
Fruit	:	Capsule completely covered by calyx

Pollination : Self pollination

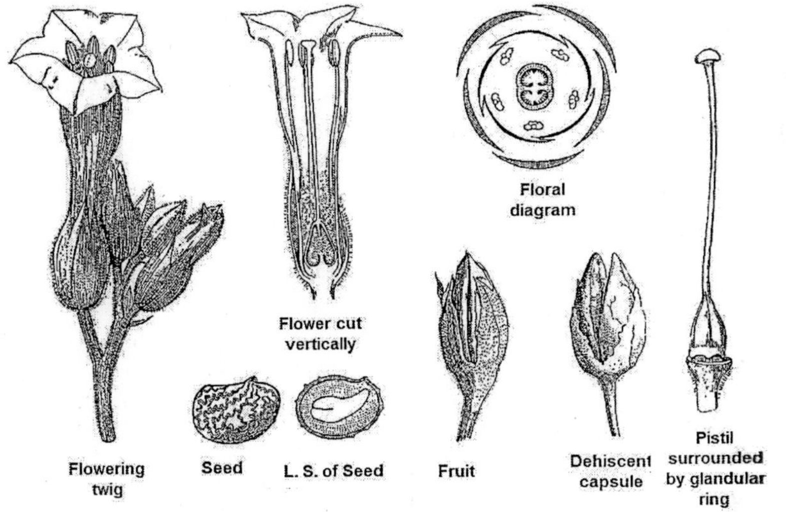

Floral formula: $Br \oplus ⚥ K_{(5)} \overset{\frown}{C_{(5)} A_5} G_{(2)}$.

Fig. 14.1. Floral structure and diagram of Nicotiana tabacum (Solanaceae)

Origin: South America

N. tobacum evolved through natural hybridization between *N. sylvestris* and *N.otophora* followed by chrcmosome doubling of the sterile F_1 hybrid. *N. rustica* evolved from natural hybridization between *N. undulate* and *N. paniculata* followed by chromosome doubling.

Wild Species of Tobacco

N. glauca	*N. bigelovii*	*N. greeneana*
N. trigonophylla	*N. plumbaginifolia*	*N. otophora*
N. clevelandii	*N. quadrivalvis*	*N. sylvestris*
N. attenuate	*N. palmeri*	*N. paniculata*
N. undulata		

Cultivated types of tobacco

***N. tobacum*:** A stout viscid annual; 1-3 m high; thick erect with few branches; leaves ovate, up to 100 cm or more in length.

***N. rustica*:** A coarse annual with thick pubescent stem and slender branches; leaves ovate or cordate; seeds are three times heavier than *N. tabacum*; generally high in nicotine content.

Topping and Desuckering

When the flower buds begin to appear, the inflorescence and top most leaves are plucked and removed. Soon after topping suckers will appear in side shoots. These are also removed. Chemicals like maleic hydrazide is used for desuckering.

Commercial types and uses of tobacco

1. **Cigarette or flue cured tobacco:** Made from finely cut tobacco, rolled into a narrow cylinder and wrapped in specially manufactured paper called cigarette paper or tissue. *N. tobacum* is widely used for this type. Also called Virginia tobacco.
2. **Cigar and cheroot type:** Mostly grown in Tamil Nadu and West Bengal. Leaf if dark brown in color, thin in texture.
3. **Beedi type:** Both *N. tobacum* and *N. rustica* are used. The leaves are thick in texture having strong flavor; green to yellowish brown in colour. Prepared by wrapping ground tobacco mixture in a special variety of leaf of *Diospyros melanoxylon*. It is the cheapest indigenous smoking tobacco.
4. **Hooka type:** Prepared by treating tobacco with Treacle a sticky dark liquid produced while sugar is being refined. Used in hookah. Both the species are used. Medium in texture and high in nicotine content.
5. **Chewing type:** Often taken with betel leaf. Thick in texture and has pungent smell.
6. **Snuff type:** Snuff is a powered tobacco being snuffed up in the nose as a stimulant or intoxicant. Thick in texture and pungent in smell.
7. **Wrapper type:** Mainly grown in West Bengal. Used for wrapping cheroot and cigar. Thin in texture.

Curing of tobacco

Curing of oxidation or dry fermentation process by which water from leaves are removed without affection the quality to tobacco. The green colour is lost and the leaves undergo chemical changes. By curing the leaves become pliable so that it can be stored and transported easily. The different curing methods are

1. **Flue curing:** Done in Brick barns where temperature and humidity are controlled. The barns are having a series of metal pipes (Flues). Heat is supplied through the pipes from outside source and sprinkling water provides humidity. Curing takes 4-6 days and the process starts with low temperature and high humidity and ends with high temperature and low humidity. Leaves have elasticity, mild aroma. Used for cigarette, hookah tobacco and also for chewing tobacco.
2. **Air curing:** Done in barns with circulating air for 6-8 weeks. Used for cigarette, hookah tobacco and also for cigars.
3. **Sun curing:** Dried on frames by hanging the leaves. Cured in direct sunlight for a period of 1 – 12 months. Used for hookah and beedi making.

4. **Fire curing:** This is the oldest method. Fire is provided in the barn during daytime and jiggery solution is sprinkled on the leaves. During night the leaves are softened. This process takes 3-4 weeks and used in cheroot, chewing and snuff types.

BOTANICAL FEATURES OF POTATO

Potato is cultivated in hilly areas of India which is used as vegetable crop

Potato: *Solanum tuberosum* $2n - 48$ **Family:** Solanaceae

Habitat	:	Tropical
Habit	:	Herbaceous annual
Root	:	Tap root with a mass of fibrous root
Stem	:	Erect herbaceous, axillary bud present below ground level develops into swollen stem tuber with leaf scar
Leaves	:	Iterruptedly imparipinnate (Imparipinnately compound leaf with interjected leaflets between the main pinnae as in tomato)
Inflorescence	:	Cyme of blue or white flowers
Flower	:	Pedicellate, Bracteate, Actinomorphic, Bisexual, Hypogynous
Calyx	:	Sepals – 5, Gamosepalous, Valvate
Corolla	:	Petals – 5, Gamopetalous, Twisted in bud, Valvate in opened flower
Androecium	:	Stamens – 5, Epipetalous
Gynoecium	:	Bicarpellary, Syncarpous, Superior ovary, Ovules – numerous in axile placentation, Capitate stigma
Fruit	:	Berry
Pollination	:	Cross Pollination

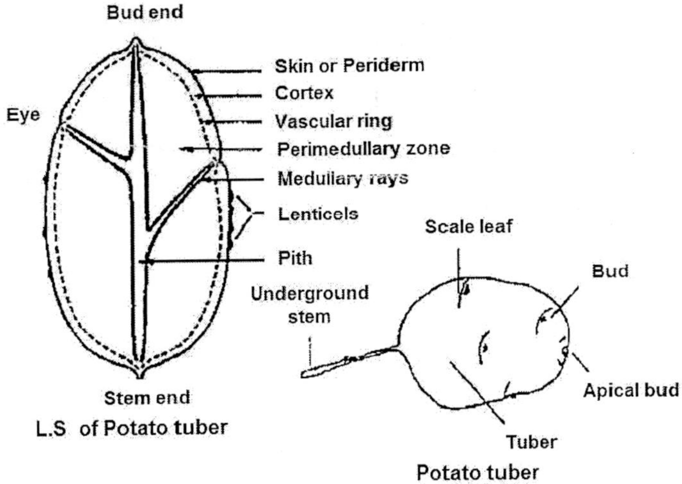

Fig. 14.2. Structure of Potato tuber

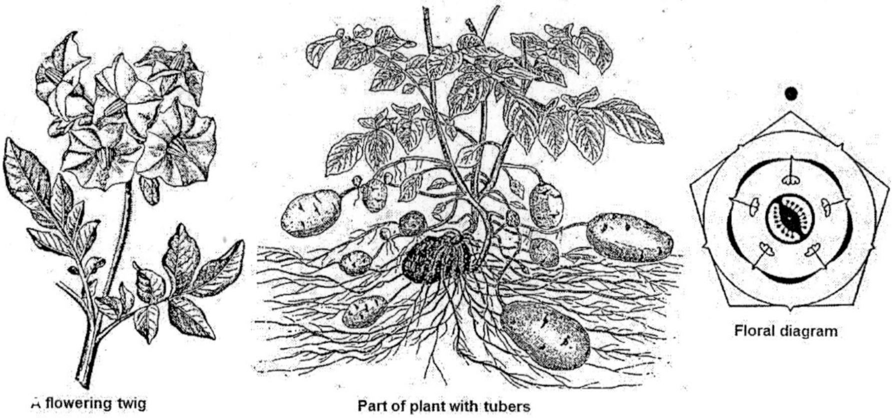

Fig. 14.3. Floral structure and diagram of Solanum tuberosum (Solanaceae)

Origin: South America

Distribution: Russia, Germany, Poland, France, USA, Britain and in India in the states of Uttar Pradesh, West Bengal, Bihar and Punjab

There are three views regarding the origin of cultivated Solanum tuberosum.

1. By natural doubling of diploid cultivated potato S. stenotomum
2. By natural crossing of diploid wild species followed by natural doubling.

S. sparsipilum *x* S. verneii

↓

F_1 F_1 -Sterile

Natural Doubling ↓

Cultivated Solanum tuberosum

3. It was also belived that cultivated *Solanum tuberosum* must have originated from *S. andigenum* followed by selection.

Classification of potato

Cultivated species are recognized which are diploids, triploids tetraploids and penta ploids.

Diploid (2n = 24)	:	*S. stenotonum* – Long duration, having dormancy
		S. phureja – Short duration no dormancy suitable for plains
Triploid (2n = 36)	:	*S. chaucha*
Tetraploid (2n = 48)	:	*S. tuberosum, S. andigenum*
Pentaploid (2n = 60)	:	*S. curtilobium – Frost resistant*

Structure of potato tuber

The potato is cultivated for it tubers; these tubers are stem structures formed by the enlargement of the tip of an underground stem. They bear 'eyes' that is, buds or groups of buds originally formed in the axils of scale-leaves. These scale leaves are short lived, and their position is marked in the mature tuber by slight ridges, the enlarged leaf scars, which form the eyebrows. It is desirable that the tuber should be of regular shape, with shallow eyes, in order to avoid waste in peeling.

The eyes are not uniformly distributed but tend to be more closely spaced towards the apical (rose end) end than towards the point (heel end) where the tuber was attached to the unthickened part of the rhizome. The leaves and other green parts of the plant consist of poisonous alkaloid 'solanine' which are not suitable for fodder. Solanine is absent in tubers, except where these are greened by exposure to light. Anatomically, the tuber shows stem structure. The great increase in diameter of the tuber is mainly due to the proliferation of the primary tissue. In transverse section, a ring of vascular bundles surrounds a irregularly circular path, tuber consists of starch in the form of large oval starch grains.

Economic importance of potato

1. Cheapest source of starchy food.
2. Human food is a variety of ways-vegetable, chips.
3. French fry, crisps, wafers.
4. Potato starch used in cloth industry. Manufacture of alcohol, glucose, lactic acid.
5. Growth media in microbiology pathological studies.
6. Tubers are used as Cattle feed.

BOTANICAL FEATURES OF CHILLI

Chilli is grown throughout the country and is used in almost all dishes. Chillies are of numerous variety and the degree of pungency varies.

Chilli: *Capsicum annuum* $2n = 24$ **Family:** Solanaceae

Habitat	:	Tropical region
Habit	:	Tender annual or perennial
Root	:	Tap root system
Stem	:	Herbaceous, angular, branched, glabrous
Leaves	:	Simple, Petiolate, Alternate, Exstipulate, Ovate, Entire, Acuminate, Glabrous
Inflorescence	:	Solitary cyme on curved peduncle
Flower	:	Pedicellate, Bracteate, Complete, Actinomorphic, Bisexual, Hypogynous
Calyx	:	Sepals -5, Gamosepalous, Persistent, Valvate aestivation

Corolla	:	Petals -5, Gamopetalous, Valvate aestivation
Androecium	:	Stamens – 5, Epipetalous
Gynoecium	:	Bicarpellary, Syncarpous, Superior ovary, Ovules – numerous in axile placentation, Blunt stigma
Fruit	:	Berry
Pollination	:	Both self and cross pollination

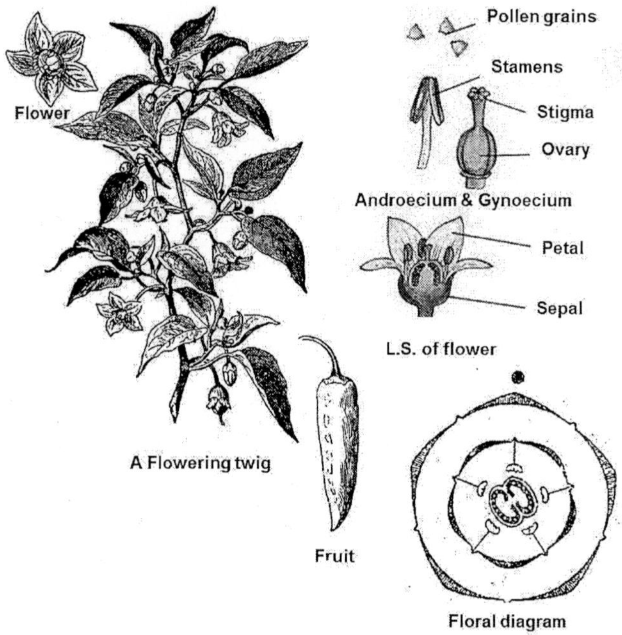

Fig. 14.4. Floral structure and diagram of Capsicum annuum

(Solanaceae)

Origin: Mexico

Cultivated species of chilli

The cultivated forms are grouped into three broad category namely,

1. Hot Peppers

1. *C. annum var. abbreviatum* – Short Pepper or wringled pepper
2. *C. annum var. conoides* – Cone Pepper
3. *C. annum var. fasiculatum* – Red Cluster Pepper
4. *C. annum var. longum* – Long Pepper. This includes Chilli, Cayenne and Long Yellow. Varieties of hot Pepper are: Hungarian Wax, Large Cherry, Long Red

Cayenne, Maule's Red-Hot, Red Chilli and Tabasco. All have red and orange-scarlet fruits when ripe. These are mainly used in sauces and flavoring.

2. Sweet or Bell Pepper

C. frutescens var. grossum – Sweet pepper which includes varieties Merrimack Wonder, Patrick Henry, Harris' Early Giant, Ruby King and Ozark Giant. The Sweet or Bell Pepper, is a popular vegetable. When the fruit is ripe it is red or yellow, but it is used as a vegetable in the green stage.

3. Ornamental chilli

C. frutescens var. cerasiforme – The Cherry Pepper, have red, purple or cream coloured fruits displayed above the rich green foliage. Although, these plants are technically perennials, they aren't worth keeping after they have fruited once. It's better to start new plants every year. Ornamental Peppers don't form good permanent houseplants.

Economic importance of chilli

1. Chilli pepper pods, which are berries, are used fresh or dried. Chillies are often dried to preserve them for long periods of time. Preserving may also be done by pickling fresh chillies.
2. Dried chillies are often ground to powders, may be reconstituted with other spices before grinding.
3. In Japanese cuisine, the leaves are cooked as greens, and also cooked in tsukukani style for preservation.
4. Chilli is also an important ingredient in almost all curries and food recipes in the country.
5. Capsaicin is a safe and effective tropical analgesic agent in the management of arthritis pain, herpes zoster related pain, diabetic neuropathy, post masectectomy pain and headaches.

Nutritional value of chilli

Chilli peppers contain more vitamin C than oranges. The amount used, however, is often very small, and most of the time chilli peppers are cooked, which reduced its vitamin C content. If the seeds are removed, they contain 50% less fiber. The proportion of various nutrients is strongly influenced by the variety. Red chilli peppers generally contain more vitamin A and vitamin C than green chilli peppers. The heat comes from capsaicin. The substances stimulates the production of saliva and activates digestion. To soften its hot taste in the mouth, have some yogurt, milk, bread, cooked rice, sugar of candy, rather than water. It also contains Capsanthin apart from Capsaicin.

BOTANICAL FEATURES OF TOMATO

Tomato plant is characterized by two types of plant:

(i) Determinate type – inflorescence occurs more frequently in almost every internode until terminal one are formed and elongation ceases at this point .In other words it is called as self – topping and the main axis terminates with a flower cluster,

(ii) Indeterminate type – inflorescence cluster occurs at every third internode and the main axis continues growing indefinitely.

Tomato: *Solanum lycopersicum* $2n = 24$ **Family:** Solanaceae

Habitat	:	Tropical region
Habit	:	Annual pubescent herb
Root	:	Tap root system
Stem	:	Branching stem, pubescent
Leaves	:	Simple, Alternate, Exstipulate, Iterruptedly imparipinnate, Reticulate
Inflorescence	:	Terminal cyme
Flower	:	Pedicellate, Bracteate, Actinomorphic, Bisexual, Pendent, Hypogynous
Calyx	:	Sepals -5, Gamosepalous, Persistent, Valvate aestivation
Corolla	:	Petals -5, Gamopetalous, Valvate aestivation
Androecium	:	Stamens – 5
Gynoecium	:	Bicarpellary, Syncarpous, Superior ovary, Ovules – numerous in axile placentation
Fruit	:	Berry
Pollination	:	Cross pollination

Origin: Tropical America

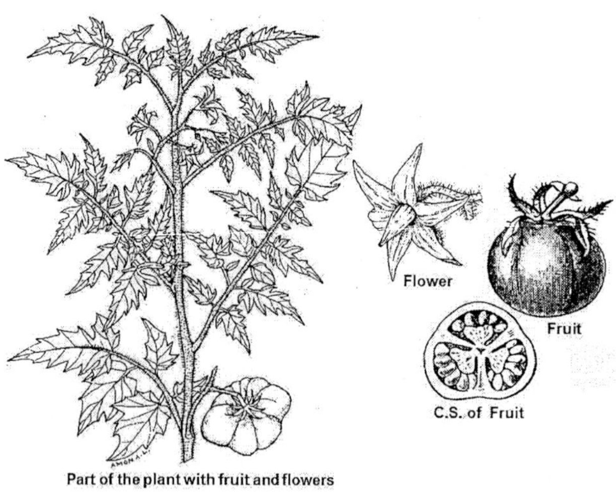

Fig. 14.5. Habit, floral and fruit structure of Lycopersicon lycopersicum (Solanaceae)

Cultivated types of tomato

1. **Common tomato:** *L. lycopersicum var. commune*
2. **Cherry tomato:** *L. lycopersicum var. cerasiforme*
3. **Potato type tomato:** *L. lycopersicum var. grandifolium*
4. **Pear type tomato:** *L. lycopersicum var. pyriforme*
5. **Upright tomato:** *L. lycopersicum var. validum*

Uses of tomato

1. The tomato fruit is consumed raw as a salad ingredient.
2. It is considered a vegetable for culinary purposes as an ingredient in many dishes.
3. It is processed into ketchup or tomato soup.
4. Unripe green tomatoes can also be breaded and fried, used to make salsa, or pickled.
5. Tomato juice is sold as a drink, and is used in cocktails such as the Bloody Mary.
6. The fruit is rich in lycopene, which may have beneficial health effects. Lycopene has a protection against oxidative damage that may lead to breast cancer, head and neck cancers and might be strongly protective against neurodegenerative diseases.

BOTANICAL FEATURES OF BRINJAL

It is important vegetable crop all over India.

Brinjal: *Solanum melongena* $2n = 24$ **Family:** Solanaceae

Habitat	:	Tropical
Habit	:	Perennial often cultivated as annual
Root	:	Tap root
Stem	:	Erect, Solid, Cylindrical, Herbaceous, Pubescent
Leaves	:	Simple, Petiolate, Alternate, Exstipulate, Reticulate, Pubescent, Coriaceous, Sinuate or shallowly lobed
Inflorescence	:	Extra axillary Solitary Cyme
Flower	:	Pedicellate, Bracteate, Actinomorphic, Bisexual, Hypogynous, Both penta and hexamerous, Violet coloured
Calyx	:	Sepals – 5 or 6, Gamosepalous, Persistent, Valvate
Corolla	:	Petals – 5 or 6, Gamopetalous, Forming a tube with five lobes, Twisted in bud, Valvate in opened flower, Violet colourededed
Androecium	:	Stamens – 5 or 6, Epipetalous, Basifixed

Gynoecium	:	Bicarpellary, Syncarpous, Superior ovary, Multi locular due to false septa
Fruit	:	Berry
Pollination	:	Often Cross pollination
Floral Formula	:	As in fig.14.6
Floral Diagram	:	As in fig.14.6
Origin	:	India

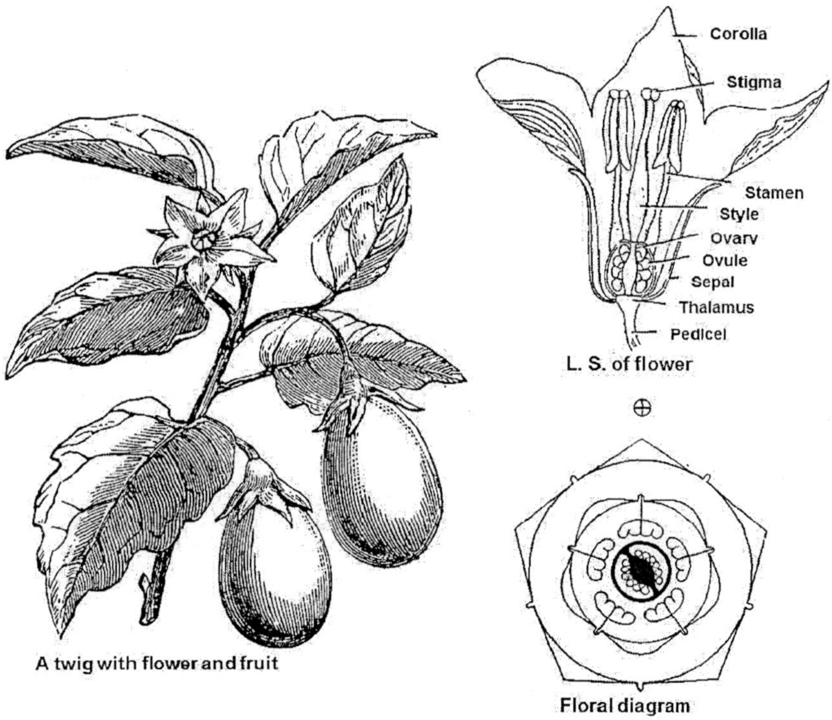

Floral formula: **Ebr**, ⊕, ⚥, $K_{(5)}$, $\widehat{C_{(5)}, A_5}$, $\underline{G_{(2)}}$

Fig. 14.6. Floral structure and diagram of Solanum melongena (Solanaceae)

Cultivated types of Brinjal

S. melongena var. esculentum : Common egg plant
S. melongena var. depressum : Dwarf eggplant
S. melongena var. serpentium : Snake eggplant
S. incanum : Local eggplant

Uses

- Used as vegetable and often described as king of vegetables.
- Egg plant is effective in the treatment of high blood cholesterol and helps to block the formation of free radicals.
- It is also source of folic acid and potassium.

LIST OF CULTIVATED CROPS UNDER SUB FAMILY MIMOSOIDEAE

Sl. No.	List of cultivated crops	Scientific name	Chromosome number	Economic Parts
	Fodder Crops			
1	Desmanthus (Valimasal)	Desmanthus virgatus	–	Whole plant
2	Subabul	Leucaena leucocephala	104	Leaves and Toppings
3	Acacia	Acacia spp	26	Leaves and Toppings

Description of sub family Mimosoideae

Plant	:	Mostly tree / shrub
Stem	:	Erect, Solid, Woody, Cylindrical, Branched
Leaf	:	Bipinnately compound leaf, Petiolate, Stipulate, Alternate, Pulvinous, Reticulate
Inflorescence	:	Axillary condenced spike (Head) with peduncle
Flower	:	Sessile, Actinomorphic, Tetra or Pentamerous, Hypogyny
Calyx	:	Sepals – 4 or 5, Gamosepalous, Valvate
Corolla	:	Petals – 4 or 5, Gamopetalous, Valvate
Androecium	:	Stamens – Numerous, Polyandrous
Gynoecium	:	Monocarpellary, Superior ovary
Fruit	:	Mostly Lomentum or Pod

Systematic position of family Mimosoideae

Kingdom : Plant kingdom
Division : Phanerogams
Subdivision : Angiosperms
Class : Dicotyledonae
Subclass : Polypetalae
Series : Calyciflorae

Order : Rosales
Family : Fabaceae
Sub family Mimosoideae

Difference between Fabaceae and Mimosoideae

Sl.No.	Traits	Fabaceae	Mimosoideae
1	Plant	Herbs	Shrub / Tree
2	Leaves	Trifoliate	Bipinnately compound
3	Flowers	Zygomorphic	Actinomorphic
4	Petals	Polypetalous, Vexillary	Gamopetalous, Valvate
5	Stamen	10 or 9+1	Numerous
6	Fruit	Pod	Lomentum

BOTANICAL FEATURES OF DESMANTHUS

Desmanthus (Velimasal or Hedge Lucerne): *Desmanthus virgatus*

Family: Mimosoideae

Habitat	:	Tropical
Habit	:	Perennial shrub
Root	:	Tap root system
Stem	:	Erect, Solid, Cylindrical, Branched, Young stem is green in colour & Older stem is woody
Leaves	:	Bipinnately compound leaf, Petiolate, Stipulate, Alternate, Pulvinous, Reticulate
Inflorescence	:	Axillary condenced spike (Head) which has bisexual, staminate and sterile flowers
Flower	:	Sessile, Bracteate, Actinomorphic, Tetra or Pentamerous, Hypogyny
Calyx	:	Sepals – 4 or 5, Gamosepalous, Valvate
Corolla	:	Petals – 4 or 5, Gamopetalous, Valvate
Androecium	:	Stamens – Numerous, Polyandrous
Gynoecium	:	Monocarpellary, Superior ovary
Fruit	:	Pod
Pollination	:	Self pollination

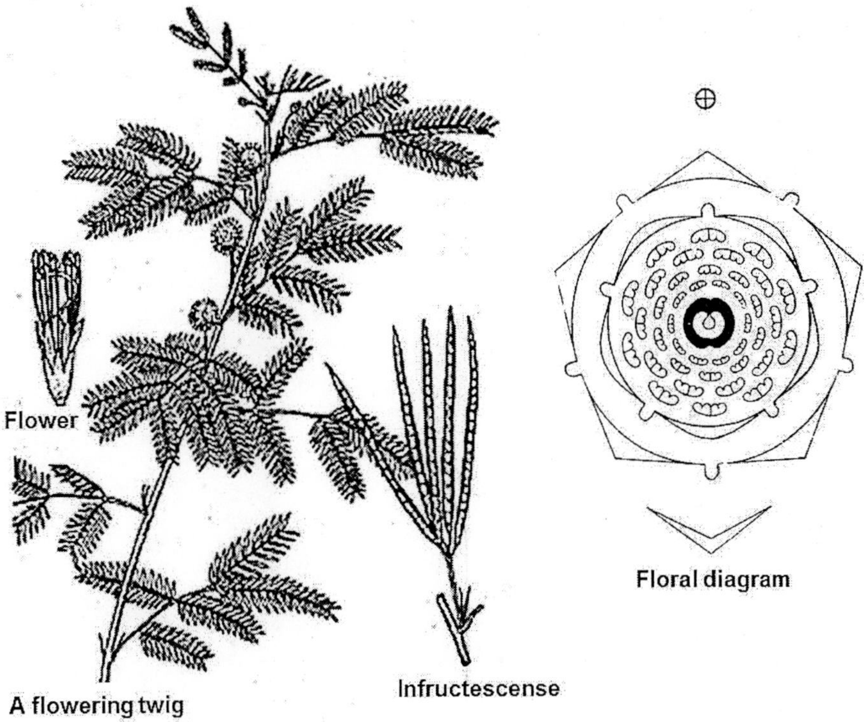

Floral formula: Br, ⊕, ⚥, $K_{(5)\ or\ (4)}$, $C_{(5)\ or\ (4)}$, A_{∞}, $\underline{G}_1$.

Fig. 14.7. Floral diagram and formula of Desmanthus virgatus (Mimosoideae)

Uses of Desmanthus

1. Grazed by herbivores in the native range.
2. Utilized in improved of permanent pastures in Autralia.
3. Grown in alley framing systems as a hedge row species in India.

BOTANICAL FEATURES OF SUBABUL

Subabul: *Leucaena leucocephala* $2n = 104$ **Family:** Mimosoideae

Habitat	:	Tropical
Habit	:	Tree / Shrub
Root	:	Tap root system
Stem	:	Erect woody
Leaves	:	Bipinnately compound leaf, Petiolate, Large gland is present in the petiole, Stipulate, Alternate, Pulvinous, Reticulate
Inflorescence	:	Axillary condenced spike

Flower	:	Sessile, Bracteate, Actinomorphic, Tetra or Pentamerous, Hypogyny
Calyx	:	Sepals – 5, Gamosepalous, Valvate
Corolla	:	Petals – 5, Gamopetalous, Five lobed, Valvate
Androecium	:	Stamens – 10, Polyandrous, Versatile
Gynoecium	:	Monocarpellary, Superior ovary
Fruit	:	Pod
Pollination	:	Cross pollination
Origin	:	Tropical America

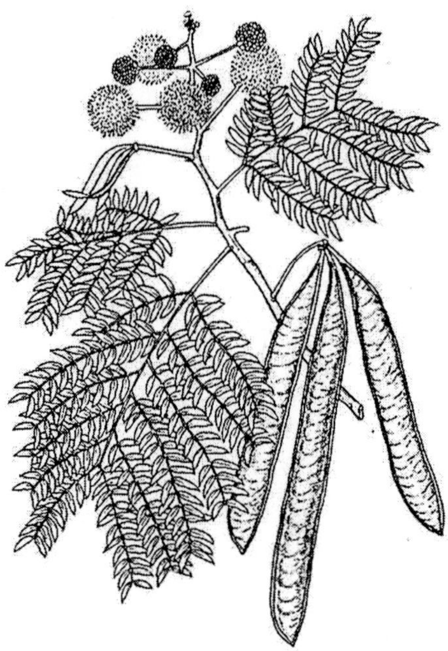

A twig with flowers and fruits

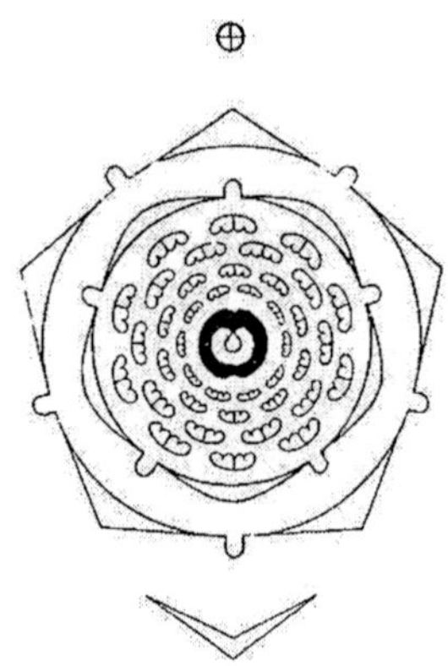

Floral diagram

Br, $\oplus$, ⚥, $K_{(5)\ or\ (4)}$, $C_{(5)\ or\ (4)}$, A_{∞}, $\underline{G}_{1}$.

Floral formula

Fig. 14.8. Floral diagram and formula of Leucaena leucocephala (Mimosoideae)

Uses of Subabul:

1. Unripe pods and seeds of all subspecies have been used by the native inhabitants of Mexico and Central America as a food or medicine since ancient times.
2. Very young shoots used as a food by villagers in Thailand.
3. Highly valued as ruminant forage.
4. Fuel wood by subsistence and semi-commercial farmers throughout southeast Asia and parts of central Asia and Africa.
5. Planted in hedge row systems with grass for cattle production.

6. Used as a shade tree over coffee and cocoa.
7. Grown in dense rows as a living fence and used to support vine crops such as pepper and passion fruit.
8. The most commonly researched species for alley farming systems.
9. Has been used as a reclamation species following mining, but no longer used due to the weed risk.

BOTANICAL FEATURES OF ACACIA

Acacia (Babool): *Acacia nilotica* $2n = 26$ **Family:** Mimosoideae

Habitat	:	Tropical
Habit	:	Perennial evergreen Tree
Root	:	Tap root
Stem	:	Erect, Cylindrical, Solid, Woody
Leaves	:	Bipinnately compound leaf, Petiolate, Alternate, Stipules modified into thorns, Pulvinus, Leaflets: Sub sessile, Oblong, Entire, Acute, Glabrous, Reticulate
Inflorescence	:	Axillary globose Spike
Flower	:	Sessile, Bracteate, Actinomorphic, Bisexual, Tetra / Pentamerous, Hypogyny
Calyx	:	Sepals – 4 or 5, Gamosepalous, Valvate
Corolla	:	Petals – 4 or 5, Gamopetalous, Valvate
Androecium	:	Stamens – Numerous, Polyandrous
Gynoecium	:	Monocarpellary, Superior ovary, Many ovules in marginal placentation, Capitate stigma
Fruit	:	Lomentum
Pollination	:	Cross pollination
Floral Formula	:	As in fig.14.9
Floral Diagram	:	As in fig.14.9

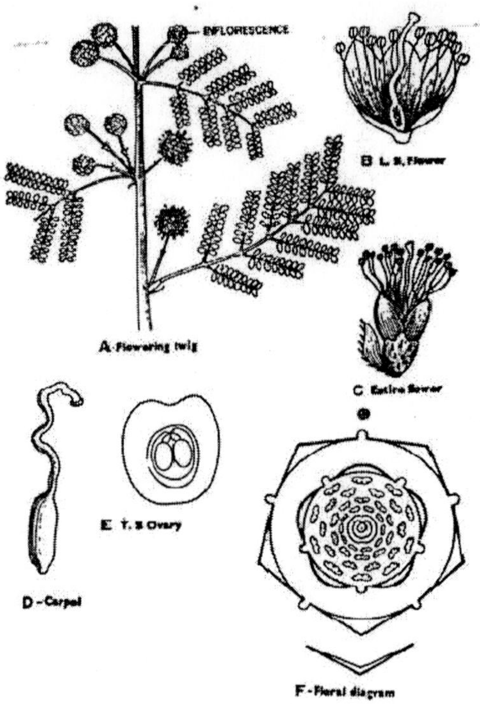

Fig. 54.1. *Acacia nilotica.*

Fig. 14.9. Floral diagram and formula of Acacia nilotica

Origin : Africa

Use of Acacia

1. The dried seeds may be toasted and salted and eaten as snack referred to as "cacalas" in Mexico.
2. The feathery shoots of Acacia pennata are used in soups, curries, omlettes and stir fries.
3. Acacia is listed as an ingredient in soft drinks, beer, energy drinks as well as in candies, mint and chewing gums.
4. Gum from Acacia senegal is called gum Arabica is abundant in dry tropical West Africa. Acacia arabica is the gum-Arabic tree of India, but yields a gum inferior to the true gum-Arabic.
5. The tanin of Acacia catechu act as astringent are used in Ayurvedic preparations especially for tooth paste.
6. Acacia root extract is used for treating premature ejaculation and rabies.
7. Acacia trees are used as ornamental trees in Gardens and roadsides and as natural fence.

DESCRIPTION OF SUB FAMILY MORACEAE

Sl. No.	Common name	Tamil name	Botanical name	2n	Economic part
1.	Mulberry	Musukottai	Morus alba	28	Leaves

Distinguishing characters of moraceae (The Mulberry family)

Habit	:	Mostly Shrubs or trees.
Root	:	Tap root system.
Stem	:	Woody, gum exudes from the stem.
Leaves	:	Petiolate, alternate, simple, stipulate with latex.
Inflorescence	:	Catkin.
Flowers	:	Unisexual, bracteates and bracteolate, actinomorphic, incomplete, hypogynous.
Tepals	:	Usually 4 in two whorls.
Stamens	:	Male flowers have 4-5 stamens and opposite to the perianth.
Gynoecium	:	Female flowers have 2 carpels, syncarpous, superior, unilocular, basal placentation.
Fruits	:	An achene, drupe or nuts.
Seed	:	Endospermic or non-endospermic.

Mulberry: *Morus alba* (2n = 28)

Family: *Moraceae*

1. BOTANICAL DESCRIPTION OF MULBERRY

- **Habit:** Shrubs or small treeswith 3-15m height.
- **Stem:** Upper portions herbaceous, lower portions woody, aerial, erect, cylindrical, branched, solid, smoothand green.
- **Leaves:** Alternate, stipulate, simple, petiolate, ovate, serrate, acute, unicostate reticulate, glabrous.
- **Inflorescence:** Catkin. Flowers unisexual, dioecious; male and female inflorescences are arranged in a spike-like catkins inflorescence

(i) **Male flower:** Ebracteate, sessile, incomplete, actinomorphic, unisexual, staminate, tetramerous and cyclic.

- **Perianth:** Tepals 4, in two whorls or 2 each, polytepalous.
- **Androecium:** Stamens 4, opposite the tepals, filaments long, dithecous, basiflxed, introrse.

- **Gynoecium:** Bicarpellary, syncarpous, ovary superior, unilocular with one pendulous ovule, style short, stigmas two.

(ii) Female flower: Ebracteate, sessile, incomplete, actinomorphic, unisexual, pistillate, tetramerous, hypogynolls and cyclic.

- **Perianth:** Tepals 4, in two whorls of 2 each, polytepalous.
- **Fruit:** Sorosis.
- **Floral formula and diagram:** As in fig. 14.9
- **Pollination:** By wind
- **Economic part:** Leaves

Chapter

15

List of Cultivated Crops, Economic parts, Chromosome Number and Family description of Cucurbitaceae and Alliaceae and Gingiberaceae

LIST OF CULTIVATED CROPS UNDER FAMILY CUCURBITACEAE

Sl. No.	List of cultivated crops	Scientific name	Chromosome number	Economic Parts
	Vegetable Crops			
1	Cucumber	*Cucumis sativus*	14	Fruit
2	Pumpkin	*Cucurbita moschata*	40	Fruit
3	Ashgourd	*Benincasa hispida*	24	Fruit
4	Melon	*Citrulus lanatus*	22	Fruit

Family description of Cucurbitaceae

Plant	:	Annual climbing herb
Stem	:	Weak with watery juice, Hairy, Pentangular
Leaf	:	Simple, Petiolate, Palmately lobed, Alternate, Exstipulate, Tendrils are present at each node
Inflorescence	:	Simple Solitary Cyme
Flower	:	Unisexual, Actinomorphic, Epigyny
Calyx	:	Sepals – 5
Corolla	:	Petals – 5
Androecium	:	Stamens – 5 or less
Gynoecium	:	Tricarpellary, Syncarpous, Inferior ovary, Ovules in parietal placentation
Fruit	:	Pepo

Systematic position of family Cucurbitaceae

Kingdom : Plant kingdom
Division : Phanerogams
Subdivision : Angiosperms

Class : Dicotyledonae
Subclass : Polypetalae
Series : Calyciflorae
Order : Passiflorales
Family : Cucurbitaceae

1. Cucumber: Cucumis sativus (2n=14)

Family: *Cucurbitaceae*

BOTANICAL FEATURES OF CUCUMBER:

Cucumber is vine crop and can be grown on the ground or on poles or trellises to hang the fruit. Cucumbers are often eaten as a vegetable but they are scientifically considered a fruit as they contain enclosed seeds and develop from a flower.

- **Origin:** South Asia

Botanical Description Of Cucumber

Habitat	:	It widely grown the tropical regions.
Habit	:	Annual, climbing herb with branching habit
Root	:	Shallow tap root system
Stem	:	The stem is 50-250 cm long robust, angular, yellowish brown and hairy. It has trailing habit, stem is slender and hollow.
Leaves	:	Leaves are simple, alternate, large with petioles. Light green to dark green in colour, palmately lobed with hairs.
Inflorescence	:	Axillary or solitary cyme
Flowers	:	Monoecious. They are yellow to deep yellow in colour. Staminate flowers are numerous borne in fascicles of 3 to 7 than the pistillate flowers are borne single. The staminate flowers can be recognized by their pedicels.
Calyx	:	Calyx is segmented on the margin of the receptacle with numerous longitudinal ribs or nerves which are clothed with long stiff hairs. White calyx tube is adnate to the ovary.
Corolla	:	Corolla is campanulate, yellow and five lobed.
Androecium	:	Male flowers are yellowish white in colour. Stamens are three, very short, free with white filaments. Anthers are adhering to the corolla tube.
Gynoecium	:	Ovary is inferior, tricarpellary, style simple with 3-5 thick stigmas.
Fruit	:	Pepo. Fruit has hard rind, pendulous on robust stalk, highly variable in shape and size usually linear cylindrical, small or large, straight or curved, green or whitish green, with yellowish white streaks at the apex. In immature fruit, the skin (pericarp) is dark green to light green in colour. On maturity, the skin colour turns to yellow.

Seeds	:	Numerous and usually flat.
Pollination	:	Cross pollination.

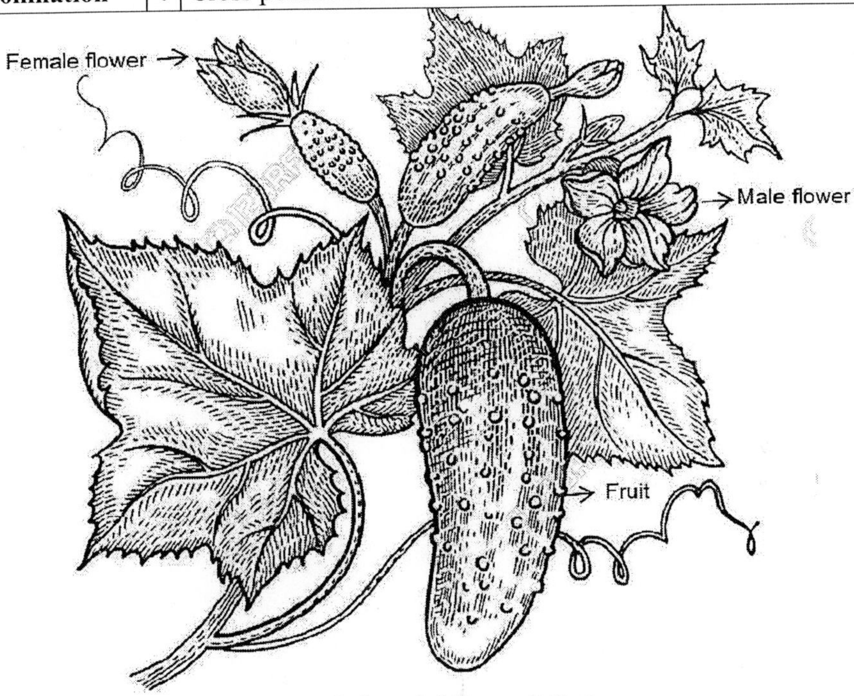

A part of plant with flower & fruit

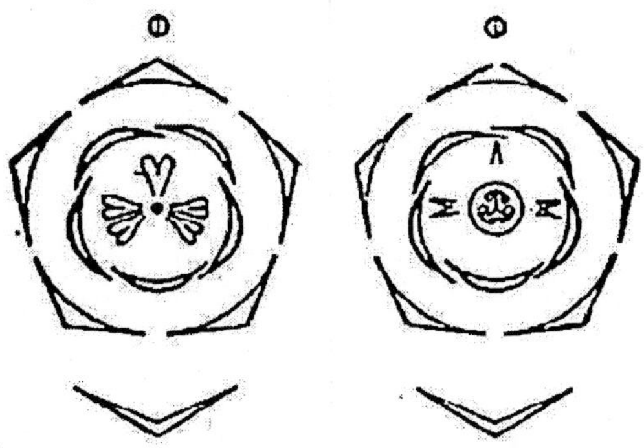

Br $\oplus$ ♂ K (5) C (5) A (2) + (2) + 1 G0 — **Floral diagram and formula of male flower**

Br $\oplus$ ♀ K (5) C (5) A 3std G (3) — **Floral diagram and formula of female flower**

Fig. 15.1. Floral structure and diagram of *Cucumis sativus* (Cucurbitaceae)

Uses of Cucumber

1. Raw fruit used as pickle, salad and soup.
2. It is also used as vegetable.

3. Small, whole, unsliced cucumbers used forGherkin, which is side dish for meals
4. Fruit slices used as face mask, skin mask for reducing skin wrinkles.
5. It is source of phosphorus, magnesium, manganese, and Vitamin A.

BOTANICAL FEATURES OF PUMPKIN

Pumpkins are cultivated in India and parts of Africa, and will withstand higher temperature. There are three cultivated species of pumpkin in India.

Curcurbita moschata (Pooshinikaai)

C. maxima (Arasaanikaai / Parangikaai)

C. pepo (Seemai Arasani / Seemai pooshini / Kumbalam)

Habitat	:	Tropical
Habit	:	Annual climbing herb, Monoecious
Root	:	Tap root
Stem	:	Weak with watery juice, Hairy, Pentangular
Leaves	:	Simple, Petiolate (Long petiole), Palmately lobed, Alternate, Exstipulate, Tendrils are present at each node, Hairy, Reticulate venation
Inflorescence	:	Axillary Solitary Cyme
Flower	:	Unisexual, Actinomorphic, Epigyny
Male flower	:	Long pedicellate, Ebracteate, Actinomorphic, Staminate, Sepals – 5, Gamosepalous, Petals – 5, Gamopetalous, Stamen – 5 (Seemingly 3), Synandrous, Ovary – rudimentary
Female flower	:	Short pedicellate, Ebracteate, Actinomorphic, Pistillate, Sepals – 5, Gamosepalous, Petals – 5, Gamopetalous, Androecium - Absent, Gynoecium: Tricarpellary, Syncarpous, Inferior ovary, Ovules in parietal placentation
Fruit	:	Pepo
Pollination	:	Cross pollination

Origin : Central America

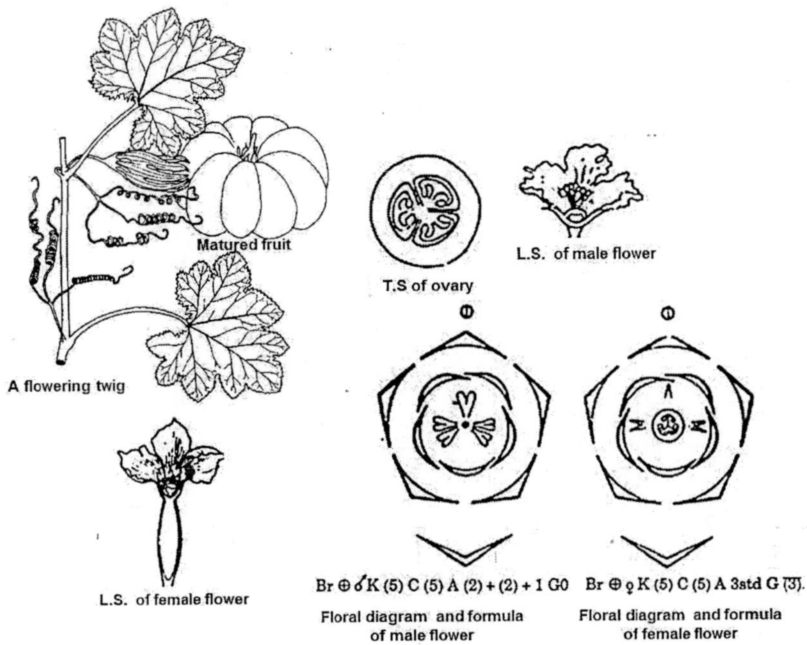

Fig. 15.2. Floral structure and diagram of Pumpkin (Cucurbitaceae)

Uses of Pumpkin

- Immature fruit – eaten as fresh vegetable.
- Mature fruit – for baking, making jams, for canning, pies, feeding cattles.
- Seed – roasted and eaten

In Asia and Africa – Leaves are cooked and eaten.

BOTANICAL FEATURES OF ASHGOURD

Ashgourd: *Benincasa hispida* $2n = 24$ **Family:** Cucurbitaceae

Habitat	:	Tropical
Habit	:	Annual and climbing herb
Root	:	Tap root
Stem	:	Shout, light green with scattered rough hairs, often five angled.
Leaves	:	Simple, Petiolate (Long petiole), Palmately lobed, Alternate, Exstipulate, Tendrils are present at each node, Hairy, Reticulate venation
Inflorescence	:	Axillary Solitary Cyme
Flower	:	Unisexual, Monoecious, Actinomorphic, Epigyny

Male flower	:	Long pedicellate, Ebracteate, Actinomorphic, Staminate, Sepals – 5, Gamosepalous, Campanulate Petals – 5, Polypetalous, Rotate Stamen – 5 (Seemingly 3) Ovary – rudimentary
Female flower	:	Short pedicellate, Ebracteate, Actinomorphic, Pistillate, Sepals – 5, Gamosepalous, Petals – 5, Polypetalous, Androecium - Absent, Gynoecium: Tricarpellary, Syncarpous, Inferior ovary, Ovules in parietal placentation
Fruit	:	Pepo
Pollination	:	Cross pollination

Origin: South America

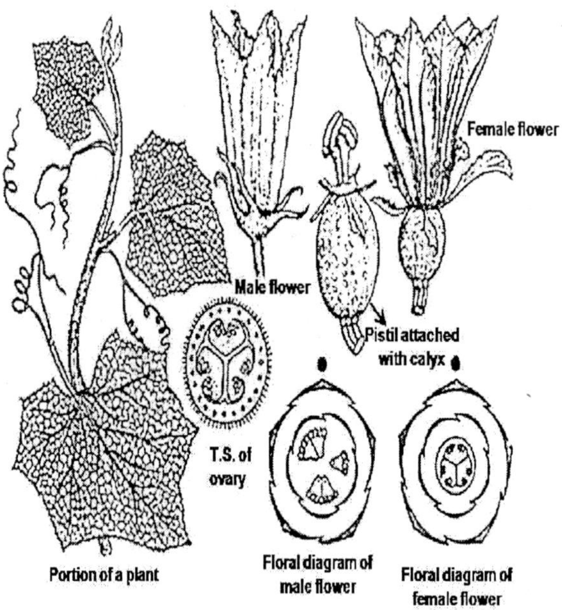

Fig. 15.3. Floral structure and diagram of *Benincasa hispida* (Cucurbitaceae)

Uses of Ashgourd

- Young fruits are boiled and eaten.
- Ripe fruits are cut into pieces and candied with sugar.

- Seeds are fried and eaten.
- Young leaf and flower are used as spinach.

BOTANICAL FEATURES OF MELON

Musk melon: *Cucumis melo* $2n - 22$ **Family:** Cucurbitaceae

Water melon: *Citrulus lanatus*

Habitat	:	Tropical
Habit	:	Annual and climbing herb
Root	:	Tap root
Stem	:	Branched, prostrate or climbing
Leaves	:	Simple, Petiolate (Long petiole), Palmately lobed, Alternate, Exstipulate, Tendrils are present at each node, Hairy, Reticulate venation
Inflorescence	:	Axillary Solitary Cyme
Flower	:	Unisexual, Monoecious, Actinomorphic, Epigyny
Male flower	:	Long pedicellate, Ebracteate, Actinomorphic, Staminate, Sepals – 5, Gamosepalous, Campanulate Petals – 5, Polypetalous, Rotate Stamen – 3, Polyandrous Ovary – rudimentary
Female flower	:	Short pedicellate, Ebracteate, Actinomorphic, Pistillate, Sepals – 5, Gamosepalous, Petals – 5, Polypetalous, Androecium - Absent, Gynoecium: Tricarpellary, Syncarpous, Inferior ovary, Ovules in parietal placentation
Fruit	:	Pepo
Pollination	:	Cross pollination

Origin : Tropical Africa

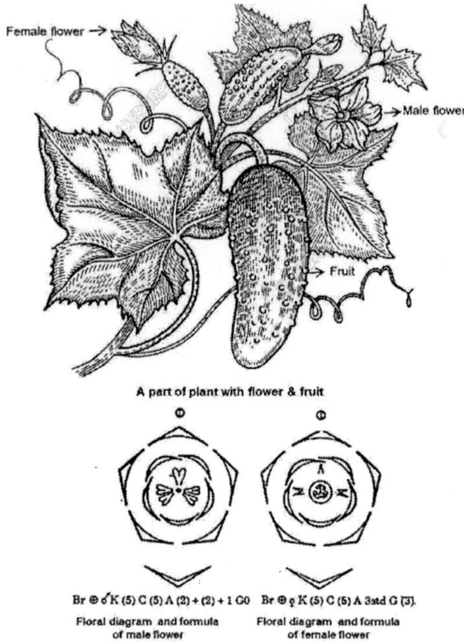

Fig. 15.4. Floral structure and diagram of *Cucumis sativus* (Cucurbitaceae)

Uses of Watermelon

- It is rich in Vitamin C, beta carotene and lycopene.
- Watermelon juice can be made into wine.
- Watermelon rinds are fed to cattle.

LIST OF CULTIVATED CROPS UNDER FAMILY ALLIACEAE (*Liliaceae*)

Sl. No.	List of cultivated crops	Scientific name	Chromosome number	Economic Parts
	Spices			
1	Onion	*Allium cepa*	16	Bulb
2	Garlic	*Allium sativum*	16	Bulbil

Family description of Alliaceae (Liliaceae)

Plant	:	Perennial herb with bulb or rhizome
Stem	:	Reduced underground stem
Leaf	:	Simple, Alternate, Radical, scapose, centric, fistular, exstipulate, Parallel venation
Inflorescence	:	Cymose Umbal

Flower	:	Pedicellate, Actinomorphic, Bisexual, Hypogynous
Perianth	:	Tepals – 6 in two whorls,
Androecium	:	Stamen – 6 in two whorls
Gynoecium	:	Tricarpellary, Syncarpous, Superior ovary, Ovules in axile placentation
Fruit	:	Loculicidal capsule

Systematic position of family Alliaceae (Liliaceae)

Kingdom	:	Plant kingdom
Division	:	Phanerogams
Subdivision	:	Angiosperms
Class	:	Monocotyledonae
Series	:	Coronarieae
Family	:	Alliaceae

BOTANICAL FEATURES OF ONION

Onion is a common condiment used in almost all dishes of curries and fries

Onion: *Allium cepa* $2n = 16$ **Family:** Alliaceae

Habitat	:	Tropical
Habit	:	Annual herb with pungent alliaceous smell cultivated for its bulb
Root	:	Fibrous root system
Stem	:	Reduced underground stem
Leaves	:	Simple, Alternate, Radical, scapose, centric, fistular, exstipulate, Parallel venation
Inflorescence	:	Cymose Umbal
Flower	:	Pedicellate, Actinomorphic, Bisexual, Hypogynous
Perianth	:	Tepals – 6 in two whorls,
Androecium	:	Stamen – 6 in two whorls
Gynoecium	:	Tricarpellary, Syncarpous, Superior ovary, Ovules in axile placentation
Fruit	:	Loculicidal capsule
Pollination	:	Cross pollination
Origin	:	West Asia

The cultivars of Allium cepa may be divided into three groups:

1. **Common Onion Group:** *Allium cepa var. cepa*. This is commercially important, shows great variation in colour and shape of the bulbs. The bulbs are large and usually single, the inflorescence typically lacks bulbils and they are almost always propagated by true seeds.
2. **Aggregatum group:** *Allium cepa var. aggregatum*. The bulbs produce many lateral bulbs or shoots which multiply freely and are usually used for propagation. The inflorescence typically lacks bulbils. Three distinct forms are recognized within the group namely;
   (a) **Potato or multiplier group:** Large bulbs with numerous lateral bulbs enclosed by outer scales. Propagated by bulb division. Seldom produce flowers
   (b) **Ever ready onion:** A perennial prolific type. Bulb and leaves narrower. Several bulbs are produced in a year
   (c) **Shallot:** Short stature, Cluster of bulbs from single planted bulb and not enclosedin the outer scales. It is widely used and propagated by bulbs
3. **Proliferum Group:** *Allium cepa var. proliferum*. It is tree onion and its inflorescence bears bulbs. Propagated by inflorescence bulb.

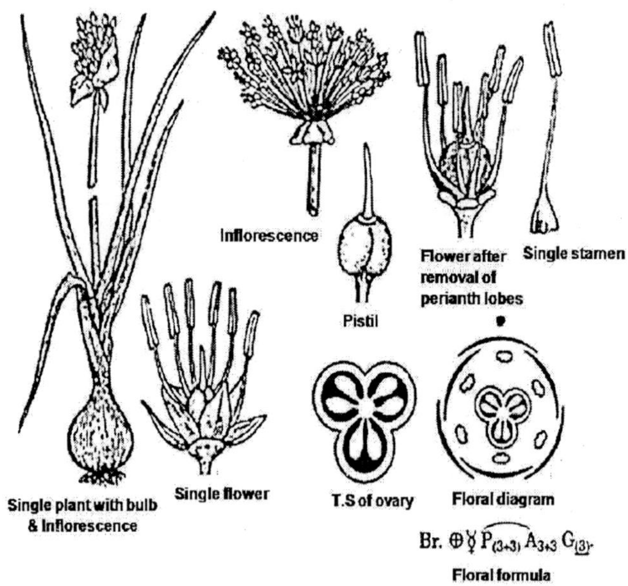

Fig. 15.5. Floral structure and floral diagram of Allium cepa (Alliaceae)

Uses of Onion

- Added in all curries as a condiment.
- Used in salad, soup and pickle.

- Dehydrated onion allow long term storage and is used in processed food.

BOTANICAL FEATURES OF GARLIC

Garlic: *Allium sativum* $2n = 16$ **Family:** Alliaceae

Garlic is used as a condiment for flavoring culinary preparation. Pungency in garlic is due to Allicin. This has got antibacterial properties. Garlic is used mainly as a strong flavouring and resembles the onion except that it has flattened solid leaves and produces a composite bulb consisting of a number of thin, sheathing, membranous leaf bases enclosing numerous small bulbs, or cloves derived from axillary buds in the axils of the leaves.

Garlic has antibacterial properties and is widely used both in intestinal disorders and for number of infectious diseases. The antibacterial activity in garlic is due to alkaloid called 'allicin' (gives pungency) which is found to inhibit growth of a number of microorganisms. Also relieves rheumatism, clears chest and improves lungs.

Habitat	:	Tropical
Habit	:	A biennial herb
Root	:	Fibrous root
Stem	:	Disc like stem
Leaves	:	Leaf blade linear, flattened solid and produces a composite bulb consisting of a number of thin, sheathing, membranous leaf bases enclosing numerous small bulbs
Inflorescence	:	Umbal
Fruit	:	Loculicidal capsule
Origin	:	Central Asia
Related species	:	*Allium longicuspis*

Cultivated types of Garlic:

1. *Allium sativum* sub *sp. ophioscorodon*: Hard necked garlic, includes porcelain garlic, rocambole garlic & purple strip garlic.
2. *Allium sativum* sub *sp. sativum*: Soft necked garlic, includes artichoke garlic, silverskin garlic & creole garlic.

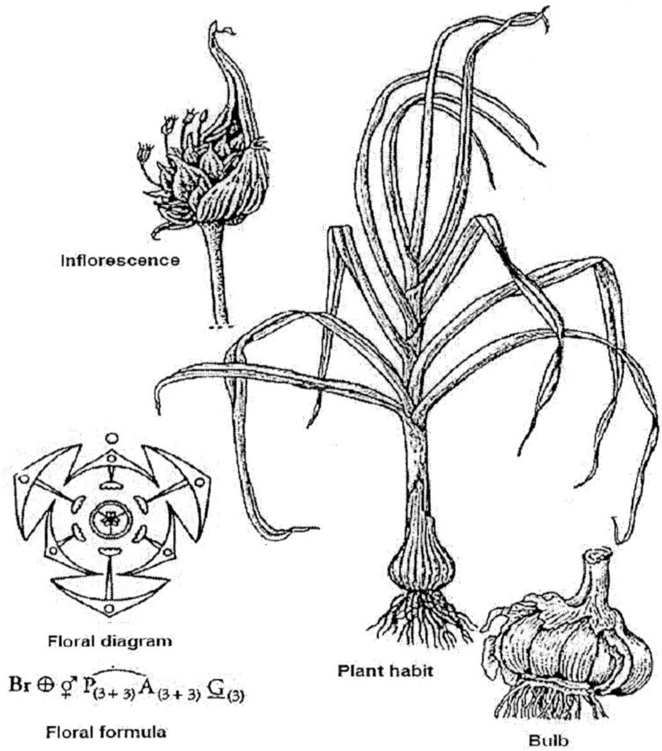

Fig. 15.6. Floral structure and floral diagram of Allium sativum (Alliaceae)

Uses

- Used as condiments in all curries.
- Dehydrated garlic powder is used in the food processing industries.
- It has been considered to be a herbal 'wonder drug'.
- It helps to lower the blood cholesterol level.
- It serves as a natural antibiotic.

LIST OF CULTIVATED CROPS UNDER FAMILY ZINGIBERACEAE

Sl. No.	List of cultivated crops	Scientific name	Chromosome number	Economic Parts
	Spices			
1	Ginger	*Zingiber officinale*	22	Rhizome

Family description of Zingiberaceae

Plant	:	Perennial herb with tuberous aromatic rhizome
Stem	:	Formed of long leaf sheath
Leaf	:	Simple, Distichous, Linear, Sheathing leaf base, glabrous, parallel venation
Inflorescence	:	Spike
Flower	:	Zygomorphic, Bisexual, Epigynous
Calyx	:	Tubular
Corolla	:	Tubular
Androecium	:	Stamens – 3 to 6,
Gynoecium	:	Tricarpellary, Syncarpous, Inferior ovary
Fruit	:	Loculicidal capsule

Systematic position of family Zingiberaceae

Kingdom	:	Plant kingdom
Division	:	Phanerogams
Subdivision	:	Angiosperms
Class	:	Monocotyledonae
Series	:	Epigynae
Family	:	Zingiberaceae

BOTANICAL FEATURES OF GINGER

Ginger contains up to 3% of fragrant essential oil whose main constituents are sesquiterpenoids and zingiberene. The pungent taste of ginger is due to gingerol which is non-volatile phenylpropanoid derived compound.

Garlic: *Zingiber officinale* $2n = 22$ **Family:** Zingiberaceae

Habitat	:	Tropical
Habit	:	Perennial herb with tuberous aromatic rhizome
Root	:	Fibrous root system
Stem	:	Formed of long leaf sheath
Leaves	:	Simple, Distichous, Linear, Sheathing leaf base, glabrous, parallel venation
Inflorescence	:	Spike
Flower	:	Zygomorphic, Bisexual, Epigynous
Calyx	:	Tubular
Corolla	:	Tubular

Androecium	:	Stamens – 3 to 6,
Gynoecium	:	Tricarpellary, Syncarpous, Inferior ovary
Fruit	:	Loculicidal capsule
Origin	:	Tropical Asia

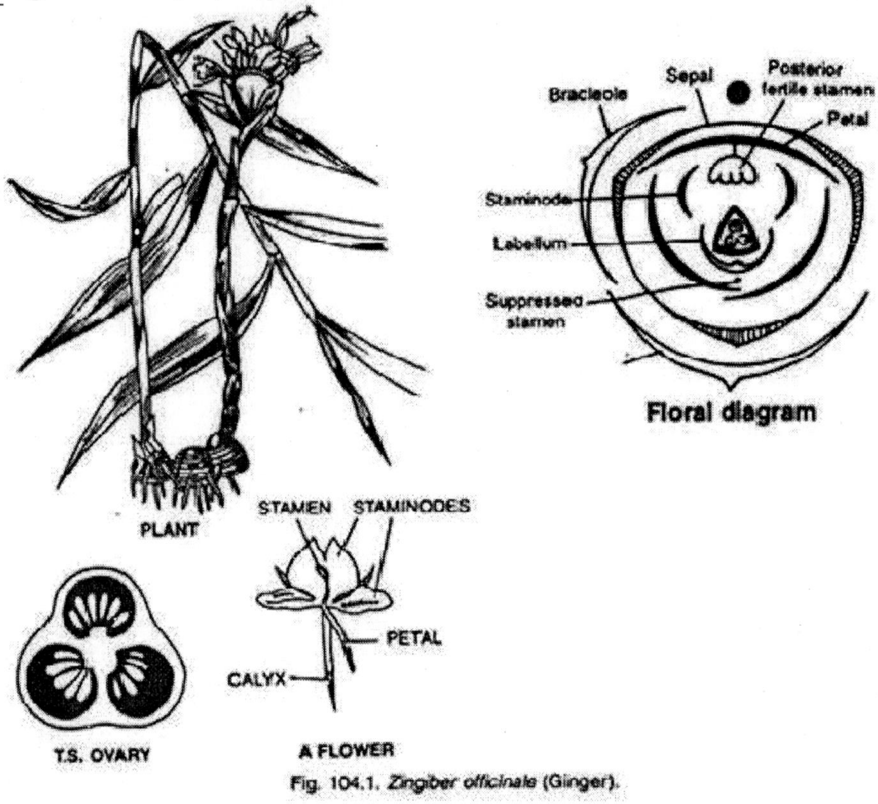

Fig. 15.7. Floral structure and floral diagram of Allium sativum (Alliaceae)

Uses

- Major use in the cookery, as flavoring agent.
- Powdered ginger : for soft drink, tinctures.
- Raw ginger : spice ginger, bread, biscuits, cakes, ginger beer and ginger wine.
- Ginger helps digestion, warms the stomach, clears the sight, and it heals joints, preparations of puddings, soups, pickles.
- Dried ginger-medicines.

Chapter
16

List of Cultivated Crops, Economic parts, Chromosome Number and Family description of Musaceae, Rubiaceae and Theaceae

LIST OF CULTIVATED CROPS UNDER FAMILY MUSACEAE (*Scitamineae*)

Sl. No.	List of cultivated crops	Scientific name	Chromosome number	Economic Parts
	Fruit crop			
1	Banana	Musa paradisiaca	33	Whole plant
	Fibre crop			
1	Manila hemp	Musa textilis	20	Leaf sheath

Family description of Musaceae (Scitamineae)

Plant	:	Giant herb looks like a tree with underground rhizome
Stem	:	Stout unbranched Pseudostem
Leaf	:	Large, Simple, with prominent mid rib and sheathing leaf base
Inflorescence	:	Mixed spadix
Flower	:	Zygomorphic, Bisexual, Epigynous
Perianth	:	Tubular
Androecium	:	Stamen – 6 in two whorls
Gynoecium	:	Tricarpellar, Syncarpous, Inferior ovary
Fruit	:	Berry

Systematic position of family Musaceae (Scitamineae)

Kingdom : Plant kingdom
Division : Phanerogams
Subdivision : Angiosperms
Class : Monocotyledonae
Series : Epigynae
Family : Musaceae

BOTANICAL FEATURES OF BANANA

Banana (Kela) is said to be the oldest cultivated tropical fruits in India. It is the second important fruit next to mango in India. Banana flavour is due to Isoamyl acetate. During ripening process, banana produces ethylene which indirectly affects the flavour.

Banana: *Musa paradisiacal* $2n = 33$ **Family:** Musaceae

Habitat	:	Tropical
Habit	:	Giant perennial herb with underground corm or sucker
Root	:	Fibrous root system
Stem	:	The real stem is underground called rhizome. The apparent, unbranched, errect and areal pseudostem is formed by the long, stiff and sheathy leaf bases which are rolled around one another to form an aerial pseudostem. The central axis that is concealed at the bottom of the pseudostem is called shaft. At the time of flowering, the shaft elongates, pierces through the pseudostem and produces an inflorescence terminally
Leaves	:	Large, Simple, Alternate, Sessile, Exstipulate, sheathing leaf base forming a pseudostem, Oblong, Entire, Glabrous, Parallel venation
Inflorescence	:	Mixed spadix covered by bracts
Flower	:	Sessile, Bracteate, Zygomorphic, Bisexual, Trimerous, Epigynous
Perianth	:	Tepals – 6, arranged in two whorls of 3 each. The three tepals of the outer whorl and the two lateral tepals of the inner whorl are fused by valvate aestivation to form 5 toothed tube like structure. The inner posterior tepal is alone free. It is distinctly broad and membranous
Androecium	:	Stamens 6, in two whorls of 3 each, Out of which 5 stamens are fertile and the inner posterior stamen is either absent or represented by a staminode
Gynoecium	:	Tricarpellary, Syncarpous, Inferior ovary, trilocular, numerous ovules on axile placentation. The style is simple and filiform, Capitate stigma
Fruit	:	An elongated fleshy berry and the seeds are not produced in cultivated varieties
Origin	:	South East Asia
Floral Formula	:	As in fig 16.1
Floral Diagram	:	As in fig 16.1

Wild species

Musa acuminate & Musa balbisiana

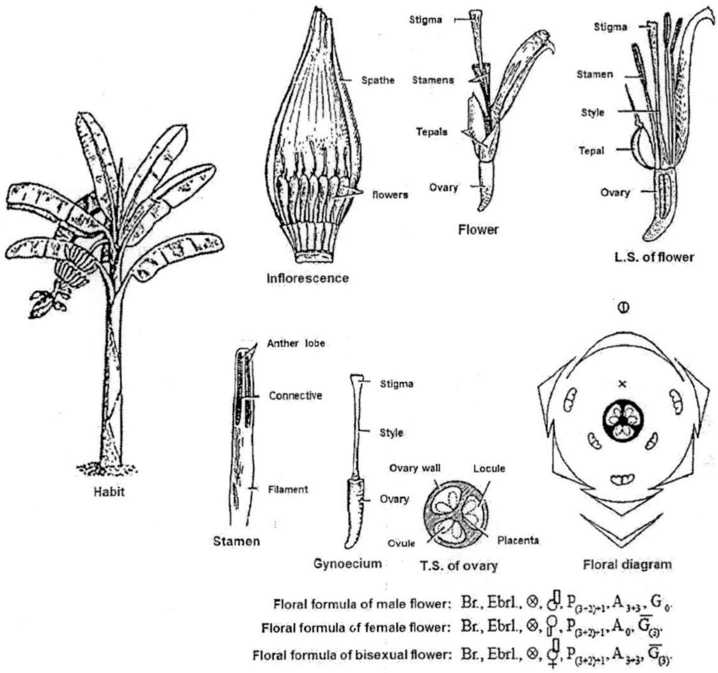

Fig. 16.1 Floral structure and diagram of Musa paradisiaca (Musaceae)

Uses

It is a good source of vitamin A and a fair source of vitamins C, B and B_2. Fruits are also rich source of minerals like magnesium, sodium, potassium and phosphorus and a fair source of calcium and iron.

BOTANICAL FEATURES OF MANILA HEMP

Manila hemp: *Musa textilis* $2n = 20$ **Fammily:** Musaceae

Habitat	:	Tropical
Habit	:	A rhizome with pseudostem formed by the leaf sheath
Root	:	Fibrous root system
Stem	:	Pseudostem
Leaves	:	Simple, Spirally arranged, Large oblong, sheath yields the leaf fibre

Inflorescence	:	Monoecious, stout, elongate, bracteates, male spikes above and female below, bracts spirally arranged, large ovate or orbicular. Bisexualflowers occur in the middle, perianth six in two series, the inner posterior tepal is large and free, the others fused, segments unequal in size and shape. Stamens six (five fertile and one staminode with rudimentary, stamen); Ovary inferior, tricarpellary, syncarpous, stigma capitates
Fruit	:	A berry
Origin	:	South East Asia

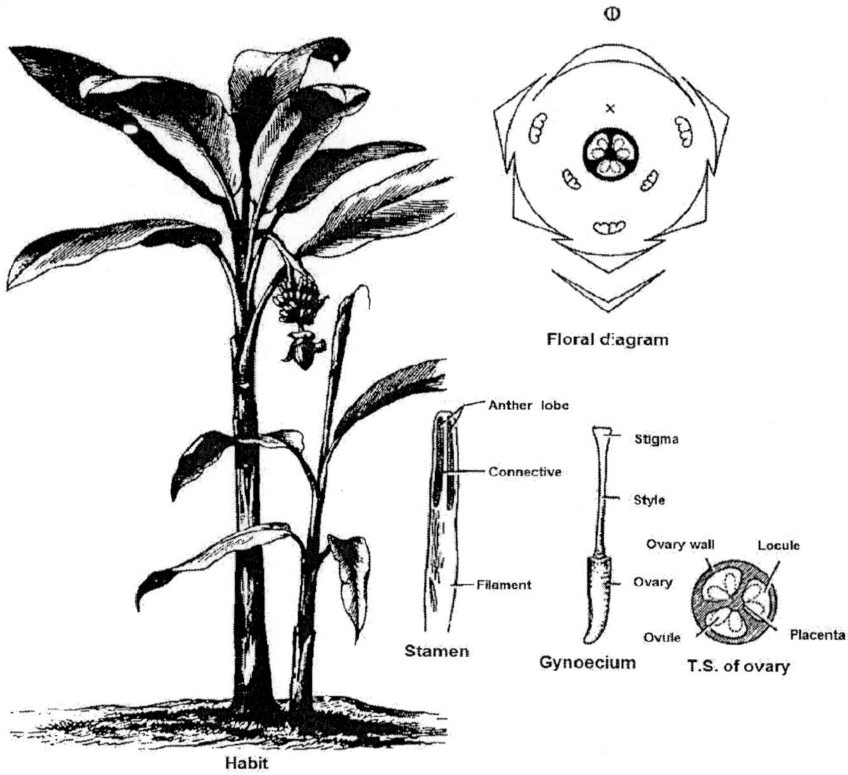

Fig.16.2. Habit, floral structure and diagram of *Musa texilis* (Musaceae)

Economic importance: Fibre from leaf.

LIST OF CULTIVATED CROPS UNDER FAMILY RUBIACEAE

Sl. No.	List of cultivated crops	Scientific name	Chromosome number	Economic Parts
	Beverages			
1	Coffee	*Coffea arabica*	44	Seeds (Beans)

Family description of Rubiaceae

Plant	:	Herbs or trees
Stem	:	Erect, Cylindrical, Solid, Branched
Leaf	:	Simple, Sub-sessile, Opposite, Inter petiolar stipules,
Inflorescence	:	Cymose
Flower	:	Actinomorphic, Bisexual, Epigynous
Calyx	:	Sepals – 5
Corolla	:	Petals – 5
Androecium	:	Stamens – 5 (As many as number of petals), Epipetalous
Gynoecium	:	Bicarpellary, Syncarpous, Inferior ovary, Ovules in axile placentation
Fruit	:	Drupe

Systematic position of family Rubiaceae

Kingdom	:	Plant kingdom
Division	:	Phanerogams
Subdivision	:	Angiosperms
Class	:	Dicotyledonae
Subclass	:	Gamopetalae
Series	:	Inferae
Order	:	Rubiales
Family	:	Rubiaceae

BOTANICAL FEATURES OF COFFEE

Coffee: *Coffea arabica* $2n = 44$ **Family:** Rubiaceae

Habitat	:	Grown in lower part of hilly areas
Habit	:	Robust evergreen tree with straight trunk
Root	:	Tap root system
Stem	:	Erect, Cylindrical, Solid, Branched
Leaves	:	Simple, Sub-sessile, Opposite, Inter petiolar stipules, Dark green leaves with shiny and waxy
Inflorescence	:	Cymose
Flower	:	Actinomorphic, Bisexual, Epigynous
Calyx	:	Sepals – 5
Corolla	:	Petals – 5

Androecium	:	Stamens – 5 (As many as number of petals), Epipetalous
Gynoecium	:	Bicarpellary, Syncarpous, Inferior ovary, Ovules in axile placentation
Fruit	:	Drupe
Origin	:	Ethiopia

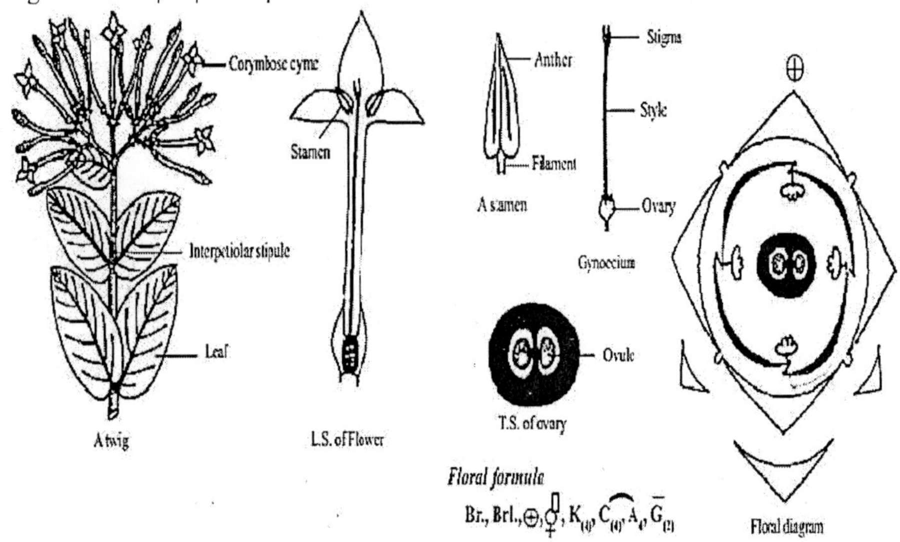

Fig.16.3. Floral structure and diagram of Coffea arabica (Rubiaceae)

Kind of Coffee

Coffee belongs to the genus Coffea which consist of nearby 25 species, of which following are of commercial importance.

1. Arabian Coffee – (*Coffea arabica*)
2. Congo Coffee, Robust Coffee – (*Coffea robusta*)
3. Liberian Coffee – (*Coffea liberica*)

'Caffeine' is an important alkaloid free from the tannin complex caffeotannic acid. The flavour of coffee is due to presence of a volatile oil caffeol.

The quality and flavour of coffee depends on number of factors such as :

- The location
- Varietal type
- The degree of ripeness of the seed
- The method of curing and drying
- The practice of roasting

Uses

Coffee has definite medicinal values acting as a diuretic and nerve stimulant, but it is also harmful if taken in large quantity. In the case of sufferers from nervous disorders, caffeine containing beverages should sparingly be given.

Instant Coffee

Powder is obtained by vaporising a strong infusion of coffee in vacuum or by the freeze drying technique.

LIST OF CULTIVATED CROPS UNDER FAMILY THEACEAE

Sl. No.	List of cultivated crops	Scientific name	Chromosome number	Economic Parts
	Beverages			
1	Tea	*Camellia sinensis*	30	Leaves

Family description of Theaceae

Plant	:	Perennial Shrub / Tree
Stem	:	Erect, Cylindrical, woody
Leaf	:	Simple, Petiolate, Exstipulate, Spiral, lanceolate, serrated, Acute apex, Reticulate venation
Inflorescence	:	Axillary solitary Cyme
Flower	:	Actinomorphic, Bisexual, Hypogynous
Calyx	:	Sepals – 5 or more, Persistent
Corolla	:	Petals – 5 or more
Androecium	:	Stamen – Numerous, Polyandrous, Epipetalous
Gynoecium	:	Tri or Pentacarpellary
Fruit	:	Berry

Systematic position of family Theaceae

Kingdom	:	Plant kingdom
Division	:	Phanerogams
Subdivision	:	Angiosperms
Class	:	Dicotyledonae
Subclass	:	Polypetalae
Series	:	Thalamiflorae
Order	:	Gutiferales
Family	:	Theaceae

BOTANICAL FEATURES OF TEA

Tea is a popular beverage equivalent to coffee. Tea is the 'Queen of beverages'. In India, main tea producing areas are Assam, West Bengal, Tamil Nadu and Kerala. Moist and warm climatic conditions are required for tea cultivation. It grows to an altitude of 2460m.

A good 150-370 cm rainfall well distributed throughout the year, absence of dry wind and freezing temperature are some of the essential requirements. Well drained acidic soil having low calcium content and rich in iron and manganese is suitable for tea cultivation.

Tea: *Camellia sinensis* $2n = 30, 45$ & 60 **Family:** Theaceae

Habitat	:	Tropical and subtropical
Habit	:	Evergreen perennial shrub
Root	:	Tap root system
Stem	:	Erect, Cylindrical, woody
Leaves	:	Simple, Petiolate, Exstipulate, Spiral, lanceolate, serrated, Acute apex, Reticulate venation
Inflorescence	:	Axillary solitary Cyme
Flower	:	Actinomorphic, Bisexual, Hypogynous
Calyx	:	Sepals – 5 or more, Persistent
Corolla	:	Petals – 5 or more
Androecium	:	Stamen – Numerous, Polyandrous, Epipetalous
Gynoecium	:	Tri or Pentacarpellary
Fruit	:	Berry
Origin	:	South East Asia

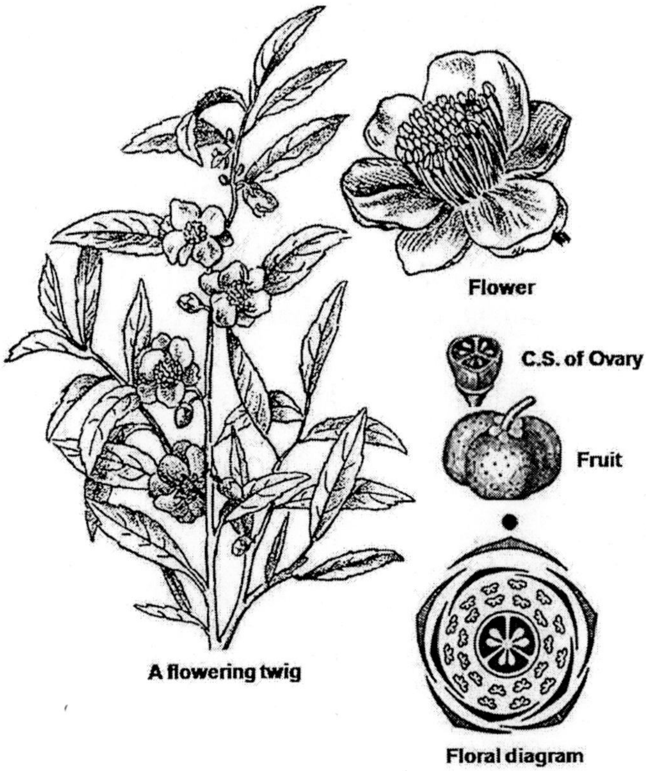

Fig. 16.4. Floral structure and diagram of Camellia sinensis (Theaceae)

Plucking of leaves

Plucking of leaves is one of the most important operation in the tea industry and it is done usually by women who carry basket on their back for collection of leaves. Constant pruning stimulates the vigorous development of new shoots called 'flushes' which are source of commercial product. The first picking is done when the plants are 3 to 4 years old and are the repeated at regular intervals. Only three or four pickings in a year are possible.

Tea Board of India

The production, marketing quality control, research etc. are controlled by the board. India and Ceylon are the main exporter of tea, where half of the world's tea is supplied from these countries, this is followed by Indonesia, China, Japan, and Taiwan. Great Britain is the great consumer of tea, followed by United States, Australia, USSR, Canada and Holland. Chinese and Japanese teas are more popular in the United States.

Grade of Tea leaves

The grade of tea leaves depends on the age of the leaves which, are given below:

1. 'Golden tips' – The youngest terminal buds rich in tannin (28%)
2. 'Orange Pekoe' (O.P.) – Smallest leaf (28% Tannin)
3. 'Pekoe' (P) – Second leaf (21% Tannin)
4. 'Pekoe Souchong' (P.S.) – The third leaf (18% Tannin)
5. 'Souchong' (S)– The fourth leaf (14% Tannin)

'Congou' (C)– The fifth and largest leaf with less tannin content.

Finally the smallest particles are graded as dusts.The flavour and quality of tea vary with the soil, climate, age of leaf, time of picking, and method of preparation.

Processing of Commercial Tea

Commercially tea may be divided into following four basic types:

1. Black tea
2. Green tea
3. Oolong tea.
4. Brick tea.

Theine is an alkaloid, identical to caffeine occurs in combination with tannins, and is only released by fermentation. The distinctive character of the beverage is due to mainly by three principle components:

- Essential oil (Ethereal, Theol oil) – For aroma and flavour.
- The akaloids fraction (Theine) – Stimulating and refreshing quality.
- Tannin (Poly phenols) – Bitterness and astringency.

QUESTION BANK: UNIT- V

BOTANICAL DESCRIPTION AND ECONOMIC USES OF TILIACEAE, PIPERACEAE, CHENOPODIACEAE, SOLANACEAE, MIMOSAE, MORACEAE, CUCURBITACEAE, ALLIACEAE, MUSACEAE, RUBIACEAE, THEACEAE

PART-A

CHOOSE THE CORRECT ANSWER — 0.5 MARK

					Key
1.	The finest tea in the world are from				
	a) Assam tea	b) Nilgris tea	c) Darjeeling tea	d) Chinese tea	c
2.	The fibre of Manila hemp is useful in				
	a) Fishing net	b) Currency paper	c) Ship's lines	d) All the three	d
3.	Examples of hill banana varieties are				
	a) Ney poovan	b) Sirumlai	c) Virupakshi	d) Both b & c	d
4.	Economic part of coffee is				
	a) Seed	b) Young leaf	c) Matured leaf	d) Fruit	a
5.	High grade coffee produced from				
	a) Coffea arabica	b) Coffea canephora	c) Coffea librica	d) All the three	a
6.	Centre of origin of coffee is				
	a) East Africa	b) Asia	c) Ethiopia	d) Uganda	c
7.	Infloresccne of banana is				
	a) Branched spadix	b) Panicle	c) Spike	d) Cyme	a
8.	Inflorescence of Coffea arabica is				
	a) Branched spadix	b) Panicle	c) Spike	d) Axillary cyme	d
9.	Which of the following alkaloid responsible for stimulating effect of tea?				
	a) Caffeine	b) Theine	c) Tannin	d) None of them	b
10.	The economic part of tea is				
	a) Seed	b) Young leaf	c) Matured leaf	d) Fruit	b
11.	The economic part of manila hemp is				
	a) Seed	b) Leaf	c) Leaf sheath	d) Fruit	c
12.	The chromosome number of banana is				
	a) 2n=24	b) 2n=30	c) 2n=33	d) 2n=40	c
13.	The economic part of garlic is				
	a) Rhizome	b) Corm	c) Bulb	d) Bulbil	d
14.	The chromosome number of onion and garlic is				Key
	a) 2n=16	b) 2n=30	c) 2n=33	d) 2n=40	a
15	The fruit type of desmanthus is				
	a) Lomentum	b) Regma	c) Achene	d) Siliqua	a

16. The centre of origin of potato is

a) South America	b) Mexico	c) Canada	d) China	a

17. The chromosome number of jute is

a) 2n=16	b) 2n=24	c) 2n=20	d) 2n=14	d

18. Fruit type sugar beet is

a) Lomentum	b) Regma	c) Utricle	d) Achene	c

19. Inflorescence of sugar beet is

a) Spadix	b) Condensed Cyme	c) Spike	d) Panicle	b

FILL IN THE BLANKS

1. Botanical name of onion is ------------------
2. Cauliflower is rich in ----------------- and ------------------------
3. Botanical name of Indian tea is -------------------------
4. Botanical name of Manila hemp is ----------------------
5. All the edible banana and plantain cultivars are hybrids of wild species of -------------and -----------------------
6. Chromosome number of garlic is -------------------
7. Subabul or savundal leaves contain amino acid is----------------------
8. King of vegetables is ----------------
9. Tomato fruit is rich in ---------------
10. Fruit type of tea is ------------------
11. Chromosome number of coffee is ---------------------
12. Coffee belongs to the ------------- family.
13. In Manila hemp the leaf sheath fibre extracted by ------------------- process.
14. Botanical name of banana is ----------------------
15. Botanical name of garlic is -----------------------
16. Cucurbits are usually -------------- pollinated crops.
17. Botanical name of subabul is ----------------------
18. Botanical name of chilli is -------------------
19. Jute contain bitter glycoside is --------------------
20. Botanical name of betel vine is----------------------

Answers keys

1. Allium cepa 2. β-carotene and β-cryptoxanthin

3. Camellia sinensis, 4. Musa textilis

5. Musa acuminata and Musa balbisiana

6. $2n = 16$, 7. Mimosin 8. Brinjal

9. Lycopene 10. Berry

11. 40
12. Rubiaceae
13. Tuxying
14. Musa paradisica
15. Allium sativum,
16. Cross
17. Leucaena leucocephala.
18. Capsicum annum
20. Corchorin
21. Piper betel

MATCH THE FOLLOWING

1.	Coffee	(a) Bulb
2.	Onion	(b) 'Caffeine'
3.	Tea	(c) Young leaves
4.	Manila hemp	(d) Plantain
5.	Banana	(e) Leaf sheath fibre

Answer key:

1. (b) 2. (a) 3. (c) 4. (e) 5. (d)

STATE TRUE OR FALSE

1. Coffee belongs to the family Theaceae.
2. Garlic contain alkaloid is Allicin.
3. Manila hemp also called as Abaca
4. Tea is the "National Drink" in India.
5. Subabul is an example of tree fodder.
6. The origin of coffea Arabica is china.
7. Cultivated banana is basically a triploid.
8. Centre of orgin of banana is Philippines.
9. Cucurbits are usually self pollinated cross.
10. Generally, cucurbitaceae family crops are monoecious.
11. Botanical name of ashgourd is Cucurbita moschata.
12. The fruit type of tomato is a berry.
13. The alkaloid Capsicin is responsible for pungency of red pepper.
14. Tomato is self pollinated crop.
15. Tobacco belongs to the solanaceae family.

Answer key:

1. False 2. True 3. True 4. True 5. True 6. False
7. True 8. True 9. False 10. True 11. False 12. True
13. True 14. False 15. True.

PART-B

1 MARK

1. Tree fodder
2. Tea picking
3. Origin of N. tabaccum
4. Curing of tobacco.
5. What is the leaf morphology of ashgourd?
6. Write cultivated types of coffee.
7. Uses of Manila hemp.
8. Utilization of jute

PART-C

2 MARK

1. Elucidate botanical description of musaceae family
2. Write the cultivated types of pumpkin and their uses
3. Write shortnotes on commercial types of tobacco
4. Elucidate the cultivated types of tea
5. Write economical importance or uses of banana
6. Write short notes on retting of jute

PART-D

5 MARK

1. Explain botanical description of Banana
2. Commercial classification and their uses of Tobacco
3. What is Curing of tobacco? Discuss the different curing method of tobacco.
4. Give complete botanical description of tobacco plant.
5. Write uses of jute. Elucidate retting of jute

Chapter 17

Summary of Botanical Description of Important Cultivated Crops

Sl. No	Common name	Botanical Name	2n	Family	inflorescence	Androecium/ No of Anthers	Gynoecium	Flower	Fruit type	Economic part	Others
I	**Cereals/ millets**										
1.	Rice	*Oryza sativa*	24	Poaceae	Terminal Panicle	6 anthers	Style bifid with plumose stigmas	Each florets consists 2 lodicules	Caryopsis	Grain	Sake-Japanese Rice wine
2.	Wheat	*Triticum aestivum*	42	Poaceae	Terminal, compound, distichous spike	3 anthers	Style 2, single ovule bifid feathery stigma Superior ovary	Each florets consists 2 lodicules	Caryopsis	Grain	Protein -Gluten
3.	Maize	*Zea mays*	20	Poaceae	Unisexual (diclinous), Monoecious, Male-tassel Female-cob	3 anthers	Single ovary with three carpels	Lodicules absent, Protandry	Caryopsis	Grain	Protein-Zein; Beverages-Bourbon whiskey
4.	Sorghum	*Sorghum bicolor*	20	Poaceae	Terminal Panicle	3 anthers	Sessile and hermaphrodite	-	Caryopsis	Grain	Beverages-Beer (Africa)
5.	Pearl millet	*Pennisetum glaucum*	14	Poaceae	Contracted panicle	3 anthers	Ovary with 2 styles	Lodicules absent, Protogyny	Caryopsis	Grain	Beverages-Beer (Africa)
	Minor millets										
1.	Finger millet (Ragi)	*Eleusine coracana*	36	Poaceae	Terminal, Sessile spikes	3 anthers	-	Lodicules 2, Hermaphrodite	Utricle	Grain	-
2.	Fox tail millet (thenai)	*Setaria italica*	18	Poaceae	Spike	3 anthers	Ovary with 2 long styles ending in plumose stigmas	-	Caryopsis	Grain	-
3.	Barnyard Millet	*Echinochola frumentosa*	36 & 54	Poaceae	Panicle	3 anthers	-	Lodicules 2	Caryopsis	Grain	-

4.	Proso millet	*Panicum milliaceum*	36 & 72	Poaceae	Slender panicle	3 anthers	2 styles plumose stigmas	Lodicules 2	Caryopsis	Grain	-
5.	Little millet (samai)	*Panicum sumatrense*	36	Poaceae	Panicle	3 anthers	Ovary with plumose stigmas	Lodicules 2	Caryopsis	Grain	-
6.	Kodo millet (Varagu)	*Paspalum scrobiculatum*	40	Poaceae	Panicle	3 anthers	plumose stigmas	Lodicules 2	Caryopsis	Grain	-
Sl. No.	Common name	*Botanical Name*	2n	Family	inflorescence	Androecium/ No of Anthers	Gynoecium	Flower	Fruit type	Economic part	Others
	Pulses										
1.	Redgram	*Cajanus cajan*	22	Papilionaceae	Terminal/ axillary raceme	Diadelphous/ (9+1) anthers	Monocarpellary, Superior ovary	Bisexual, zygomorphic	Pod	Seed	-
2.	Bengal gram	*Cicer arietinum*	16	Papilionaceae	Axillary solitary raceme	Diadelphous/ (9+1) anthers	Monocarpellary, Superior ovary	Bisexual, zygomorphic	Pod	Seed	-
3.	Soybean	*Glycine max*	40	Papilionaceae	Axillary raceme	Monoadelphous (10 stamens)	Hairy sessile	-	Pod	Seed	29-50% protein
4.	Black gram	*Vigna mungo*	22 & 24	Papilionaceae	Axillary raceme	Diadelphous/ (9+1) anthers	Superior ovary	-	Pod	Seed	23% protein
5.	Green gram	*Vigna radiata*	22	Papilionaceae	Axillary raceme	Diadelphous/ (9+1) anthers	Superior ovary	Complete, zygomorphic	Pod	Seed	-

6.	Cow pea	*Vigan ung-iculata*	22 & 24	Papili-onaceae	Axillary raceme	Diadel-phous/ (9+1) anthers	Ovary sessile, Hypoynous	Pedicellate, zy gomorphic	Pod	Seed	-
7.	Lab lab	*Dolichos lablab*	22 & 24	Papili-onaceae	Lax fascicled	Diadel-phous/ (9+1) anthers	Style filiform & glabrous	Short pedicle	Pod	Seed	-
8.	Horse gram	*Macro-tyloma uniflorum*	24	Papili-onaceae	Axillary raceme	Diadel-phous/ (9+1) anthers	Style filiform & glabrous	Zygomor-phic	Pod	Seed	-
Forage legumes											
1.	Lucern	*Medicago sativa*	16, 32 & 64	Papili-onaceae	Axillary raceme	Diadel-phous/ (9+1) anthers	Superior ovary	-	Pod	Leaves & toppings	Queen of fodder
2.	Stylosan-thus	*Stylosanthus hamata*	40	Papili-onaceae	Axillary raceme	Diadel-phous/ (9+1) anthers	Superior ovary	Standard	Pod	Leaves & toppings	-
3.	Clitoria	*Clitoria ternatea*	16	Papili-onaceae	Axillary raceme	Diadel-phous/ (9+1) anthers	Superior ovary	Pedicle	Pod	Leaves & toppings	-
4.	Agathi	*Sesbania grandiflo-rum*	24	Papili-onaceae	Axillary raceme	Diadel-phous/ (9+1) anthers	Superior ovary	-	Pod	Leaves & toppings	-
	Green manure crop										
1.	Sunhemp	*Crotalarial juncea*	16	Papili-onaceae	Terminal open raceme	Monoadel-phous (10 stamens)	-	Pedicellate	Pod	Whole plant	Stem used bast fibre

2.	Daincha	*Sesbania aculeate*	12 & 24	Papili-onaceae	Axillary raceme	Diadel-phous/ (9+1) anthers	Superior ovary	Papilina-ceous	Pod	Whole plant	-
3.	Manila agathi	*Sesbania rostrata*	12	Papili-onaceae	Axillary raceme	Diadel-phous/ (9+1) anthers	Superior ovary	Papilina-ceous	Pod	Whole plant	-
	Sugar crop										-
1.	Sugarcane	*Saccharum officinarum*	80	Poaceae	Arrow	3 anthers	2 styles with purplish feath-ery stigma	-	Caryopsis	Stem	-
Sl. No	Common name	*Botanical Name*	2n	Family	inflorescence	Androe-cium/ No of An-thers	Gynoecium	Flower	Fruit type	Economic part	Others
	Oil seeds										
1.	Groundnut	*Arachis hypogea*	40	Papili-onaceae	Spikes	Monoadel-phous (10 stamens)	Monocarpel-lary, Long filiform style	Sessile, bisexual, zy-gomorphic	Pod	Seed	-
2.	Gingelly	*Sesamum indicum*	26	Pedalia-ceae	Solitary axil-lary cyme	Stamens 4, didynamous	Bicarpellary, Hypogynous	Sessile, syncarpous	Locu-licidal capsule	Seed	-
3.	Sunflower	*Helianthus annus*	34	Astera-ceae	Head or Capitulum	Syngene-cious, 5 anthers	Bicarpellary, Epigynous	Sessile, Ray & disc floret	Achene	Seed	-
4.	Safflower	*Carthamus tinctorius*	24	Astera-ceae	Head or Capitulum	5 anthers	-	Ray & disc floret	Achene	Seed	-
5.	Mustard	*Brassica junceae*	36	Brassi-caceae	Raceme	6 anthers	Bicarpellary, Hypogynous	Complete, actinomor-phic	Sliqua	Seed	-

6.	Castor	*Ricinus communis*	20	Euphorbiaceae	Terminal panicle cyme	5 anthers	Hypogynous	Actinomorphic	Schizocarpic capsule	Seed	-
7.	Coconut	*Coccous nucifera*	32	Arecaceae	Compound spike or spadix	6 stamens	Tricarpellary, syncarpous	Monoecious	Drupe	Endosperm	-
8.	Oil palm	*Elaeis guineensis*	32	Arecaceae	Compound spike or spadix	6 stamens	Tricarpellary	Monoecious	Drupe	Mesocarp & kernel	-
	Fibre crops										
1.	Cotton	*Gossypium spp.*	26,52	Malvaceae	Axillary solitary cyme	Reniform anthers	Syncarpous	Complete, acitnomorphic	Loculicidal capsule	Lint	Seed fibre
2.	Mesta	*Hibiscus cannabinus*	36	Malvaceae	Axillary solitary cyme	Numerous stamens	Pentacarpellary	-	Globose capsule	Stem	Stem fibre
3.	Jute	*Corchorus spp.*	14	Tiliaceae	Solitary cyme	Stamens 10 to many	Pentacarpellary, Axile placentation	Complete, acitnomorphic	Globose long capsule	Stem	Stem fibre
4.	Manila hemp	*Musa textilis*	20	Musaceae	Long drooping spike	5stamens	Inferior ovary, trilocular	Zygomorphic	Berry	Leaf sheath	Leaf sheath fibre
	Starch crops										
1	Tapioca	*Manihot uilissima*	36	Euphorbiaceae	Axillary raceme	10 stamens	Tricarpellary	Unisexual flower	Capsule	Root tuber	-
2.	Potato	*Solanus tuberosum*	48	Solanaceae	Solitary cyme	6 stamens	Pistil with several loculi	-	Berry	Stem tubers	-

Chapter

18

Model Questions for Midsemester Examination

1. PBG. 101- Introduction to Agricultural Botany (1+ 1)

I B.Sc.,[Ag] / BSc Hons Agri) I semester

Model Question-1
MID-SEMESTER EXAMINATION

Max marks: 20

PART-A

I. Choose the correct answer [20 x 0.5 = 10]

A1. The binomial nomenclature includes

a) Generic name b) Species name c) Race d) botha a & b

A2. The number of anthers in rice

a) 3 b) 6 c) 4 d) 5

A3. Male inflorescence of maize is called

a) Cob b) Tassel c) Arrow d) Spike

A4. When both male and female flowers are present in the same plant is called

a) Monoecious b) Dioecious c) Polygamous d) None of them

A5. A flower in which the overy is inferior and other floral organs arise above it condition is

a) Hypogynous b) Epigynous c) Perigynous d) All of them

II. Fill in the blanks

A6. The organization that approves the binomial nomenclature of plant is ---------------

A7. The condition of flowers, ovule mature before the anther is known as ----------------

A8. The type of venation in rice leaf is ----------------------

A9. The botanical name of finger millet (ragi) is ------------------

A10. The chromosome number of sugarcane is --------------

III. Match the following

A11. Protogyny - Potato
A12. Utricle - Pearlmillet
A13. 2n =24 - Ragi
A14. Bourbon whisky - Rice
A15. Tuber - Maize

III. State True or False

A16. Wheat floret has six stamens
A17. Pearl millet grains are rich in tryptophan and cystine
A18. Finger millet is an example of famine food
A19. The leaf margin that is toothed termed as serrated
A20. In general, the dicot plants have parallel venation

PART-B

I. Answer any FIVE questions only [5 x 2 = 10]

1. Write out line of Bentham and Hooker plant classification
2. Enumerate the classification of maize grains based on grain structure
3. Explain Nobilization of sugarcane
4. Elucidate the economic importance of rice
5. Write orgin of hexaploid wheat
6. Writhe characteristics feature of poaceae family.

2. PBG. 101- Introduction to Agricultural Botany (1+ 1)

I B.Sc.,[Ag] / BSc Hons Agri) I semester | **Model question-2 MID-SEMESTER EXAMINATION** | **Max marks: 20**

PART-A

I. Choose the correct answer: **[6 x 0.5 = 3]**

1. The chromosome number for Rice is
   a) 2n : 20 b) 2n: 24 c) 2n: 26 d) 2n: 28
2. The number of anthers in rice
   a) 3 b) 6 c) 4 d) 5
3. Botanical name of Maize is
   a) Zea mays b) Oryza sativa c) Triticum aestivum d) Sorghum bicolor
4. Wheat revolution in India started by
   a) M.S.Swaminathan b) Norman Borlaug c) C.A. Barber d)T.S. Venkatraman
5. Nobilization of Sugarcane in India started at
   a) Coimbatore b) Hyderabad c) New Delhi d) Cochin
6. The alcoholic drinks prepared from Maize grain is
   a) Rum b) Bourbon Whiskey c) Beer d) Vodka

PART-B

II. Answer any THREE questions only **[3 x 1 = 3]**

1. What is Phyllotaxy and Venation?
2. Write short notes on Sorghum poisoning.
3. What is mean by small millets? List out the small millets.
4. Uses of Guinea grass.

PART-C

III. Answer any TWO questions only **[2 x 2 = 4]**

1. Origin of Tetraploid and Hexaploid wheat
2. Nobilization of Sugarcane
3. Uses of Sugarcane

PART-D

IV. Answer any TWO questions only **[2 x 5 = 10]**

1. Write an outline of Bentham and Hooker classification and their merits and demerits
2. Explain the economical importance of Rice and Maize
3. Elucidate the classification of Maize based on grain structure

3. PBG 101 Introduction to Agricultural Botany (1+ 1)

I B.Sc.,[Ag] / BSc Hons Agri) I semester

Model question: 3
MID-SEMESTER EXAMINATION

Max. Marks : 20

PART-A

I. Choose the correct answer **[6 x 0.5 = 3]**

1. Bentham and Hooker plant classification is based on
   a) Artificial system b) Natural system c) Phylogenetic system d) None
2. The number of anthers in Pearl millet is
   a) 3 b) 6 c) 4 d) 5
3. Botanical name of Rice is
   a) Zea mays b) Oryza sativa c) T. aestivum d)S. bicolor
4. Cereal crop grains are rich in
   a) Carbohydrate b) Protein c) Fat d) Fibre
5. Which crop having high Hydrocyanic acid (HCN) content at the early growth stage
   a) Wheat b) Sorghum c) Ragi d) Maize
6. The alcoholic drinks prepared from Sugarcane molasses is
   a) Rum b) Whiskey c) Beer d) Vodka

PART-B

II. Answer any three questions only **[3 x 1 = 3]**

1. What is mean by protandry and protogyny with examples?
2. Chemical composition of Rice.
3. Write short notes on barboiling of rice and their merits.
4. Uses of Ragi.

PART-C

III. Answer any two questions only **[2 x 2 = 4]**

1. Classification of sorghum based on coverage of glume on the grain.
2. Nobilization of Sugarcane
3. Industrial uses of Maize

PART-D

IV. Answer any two questions only **[2 x 5 = 10]**

1. Write an outline of Bentham and Hooker classification and their merits and demerits
2. a) Economical importance of Wheat (2.5 mark)
   b) Economical importance of Sugarcane (2.5 mark)
3. a). Classification of rice based on the quality traits (2.5 mark)
   b). Utilization of pearl millet and sorghum (2.5 mark)

Chapter

19

Model Question for Final Theory Examination

I B.Sc.,[Ag] / BSc Hons Agri) I semester

Model question-1
Final Theory Examination

Time : 2 ½ hours
Max. Marks : 40

PBG 101 Introduction to Agricultural Botany (1+ 1)
PART-A

(Answer All questions) **[20 x 0.5= 10]**

Choose the correct answer

A1. The chromosome number for Rice is

a) 2n : 20 b) 2n: 24 c) 2n: 26 d) 2n: 28

A2. Cereal crop grains are rich in

a) Carbohydrate b) Protein c) Fat d) Fibre

A3. Male inflorescence of Maize is called

a) Cob b) Tassel c) Silk d) Ovary

A4. Centre of origin of Sorghum and Coffee is

a) East Africa b) Asia c) Ethiopia d) Uganda

A5. The protein and oil content of soybean is

a) 40 & 20 % b) 20 & 30 % c) 20 & 40 % d) 20 & 50 %

Fill in the Blanks

A6. Botanical name of Soybean is ----------------------

A7. Poison present in Sorghum plant is ---------------------

A8. The condition of flowers, anthers mature the before the ovule is known as -------------------

A9. Botanical name of upland cotton is ----------------------------

A10. Expansion of ICBN--------------------------------

Match the following

A11. Stem modification - Wheat

A12. Gluten - Maize

A13. Zein - Onion

A14. Corchorin - Plantain

A15. Banana - Jute

State true or false

A16. Rice stem is also known as culm

A17. Coffee belongs to the family Theaceae

A18. Fruit type of Rice is Utricle

A19. Bourborn Whisky obtained from maize grains

A20. Jute belongs to the family Tilliaceae

PART-B

(Answer any Five questions) [5 x 1=5]

B1. Differentiate monocots and dicots

B2. What is Phyllotaxy? Name the types of phyllotaxy

B3. What is mean by Protandry and Protogyny? Give example for each

B4. List out the legume forages with botanical names

B5. What is Ginning percentage?

B6. Tea picking

PART-C

(Answer any Five questions) [5 x 2=10]

C1. Nobilization of Sugarcane

C2. Differentiate Desi and Kabuli types of Bengal gram

C3. Pod or Peg development in Groundnut

C4. List out two fibre crops, their scientific name, family and economic parts used as fibre

C5. Differentiate the family Poaceae and Fabaceae

C6. Botanical description of Musaceae family

PART-D

(Answer any Three questions) [3 x 5=15]

D1. What are the characteristics features of poaceae family? Write uses of rice

D2. Economic importance of Maize

D3. What is tapping for coconut toddy? Write uses of coconut

D4. Commercial classification and their uses of Tobacco

I B.Sc.,[Ag] / BSc Hons Agri) I semester

Model question-2
Final Theory Examination

Time : 2 ½ hours
Max. Marks : 40

PBG 101 Introduction to Agricultural Botany (1+ 1)

PART-A

(Answer All questions) **[20 x 0.5 = 10]**

Choose the correct answer

A1. The number of anthers in rice

a) 3 b) 6 c) 4 d) 5

A2. Pulses crop seeds are rich in

a) Carbohydrate b) Protein c) Fat d) Fibre

A3. Female inflorescence of Maize is called

a) Cob b) Tassel c) Silk d) Ovary

A4. Botanically sugarcane inflorescence is called as

a) Spike b) Arrow c) Spadix d) Tassel

A5. Biodiesel yielding plant

a) Groundnut b) Jatropha c) Castor d) Gingelly

Fill in the Blanks

A6. Botanical name of Regram is ---------------------

A7. Poison present in Sorghum plant is --------------------

A8. The condition of flowers, ovule mature the before the anther is known as -------
----

A9. Botanical nameof Banana is --------------------------

A10. "-------------------" is the bitter glycoside presented in Corchorus capsularis

Match the following

A11. Redgram - Leaf sheath fibre

A12. Manila hemp - Fabaceae

A13. HCN - Capitulum

A14. Sunflower - Young leaves

A15. Tea - Sorghum

State true or false

A16. Bentham and Hooker plant classification is based on artificial system

A17. Jute belongs to the family Tilliaceae

A18. The alcoholic drinks Rum is prepared from Sugarcane molasses

A19. Inflorescence of Sunflower is Capitulum

A20. Blackgram belongs to Fabaceae family

PART-B

(Answer any Fivequestions) **[5 x 1 = 5]**

B1. Draw a general morphology of dicot plant with parts

B2. List out the small or minor millets and write their botanical names.

B3. Origin of G. hirsutum

B4. Write Head or Capitulum inflorescence

B5. Cumbu Napier hybrid

B6. Write Inflorescence of coconut

PART-C

(Answer any Five questions) **[5 x 2 = 10]**

C1. Uses of Sugarcane

C2. Differentiate Black gram and Green gram

C3. Utilization of Groundnut

C4. List out two fibre crops, their scientific name, family and uses

C5. Extraction of sugar from Palmyra palm

C6. Write short notes on retting of jute

PART-D

(Answer any Three questions) **[3 x 5 = 15]**

D1. What are the characteristics features of Musaceae family? Write uses of Banana

D2. Pod development in ground nut with diagram and Write utilization of ground nut

D3. List out the legume forages and their uses

D4. Classification of Maize based on the grain structure with suitable diagrams

I B.Sc.,[Ag] / BSc Hons Agri) I semester | **Model question-3 Final Theory Examination** | **Time : 2 ½ hours Max. Marks : 40**

PBG 101 Introduction to Agricultural Botany (1+ 1)

PART-A

(Answer All questions) **[20 x 0.5 = 10]**

Choose the correct answer

A1. Which of the following havirg monoadelphous stamens

a) Black gram b) Red gram c) Lab lab d) Soybean

A2. The synthetic textile fibre derived from protein of groundnut is

a) Ardil b) Arachidill c) Fibril d) Rilson

A3. Economic part of tea is

a) Seed b) Young leaf c) Matured leaf d) Fruit

A4. Structural or leaf fibre obtained from

a) Manila hemp b) Sun hemp c) Mesta d) Cotton

A5. High grade coffee produced from

a) Coffea arabica b) Coffea canephora c) Coffea librica d) All the three

Fill in the Blanks

A6. Onion and Ginger are formed by ------------------modification

A7. Tobacco belongs to ---------------- family

A8. Chromosome number of Maize is--------------------

A9. Botanical name of Pearl millet is ---------------------------

A10. Botanical name of Sunflower is ---------------------------

Match the following

A11. Maize - Terminal raceme

A12. Sugarcane - Utricle

A13. Ragi - Panicle

A14. Rice - Arrow

A15. Redgram - Tassel

State true or false

A16. Economic part of sugarcane is stem

A17. Tea belongs to the family Theaceae

A18. Fruit type of Ragi is caryopsis

A19. Inflorescence of Sugarcane is Arrow

A20. Groundnut belongs to Malvaceae family

PART-B

(Answer any Five questions) **[5 x 1=5]**

B1. Define Phyllotaxy

B2. Define Aestivation

B3. List out the wild species used for cotton improvement

B4. Origin of N. tabaccum

B5. Tree fodder

B6. Inflorescence of Sunflower

PART-C

(Answer any Five questions) **[5 x 2=10]**

C1. Economic importance of Banana

C2. Differentiate Black gram and Green gram

C3. Write botanical description of Rice

C4. List out two oil seed crops, their scientific name, family and uses

C5. Illustrate the genetic relationship (U triangle) of oilseed Brassica by Nagaharu U (1935)

C6. Industrial uses of Maize

PART-D

(Answer any Three questions) **[3 x 5=15]**

D1. Write an outline of Bentham and Hooker classification and their merits and demerits

D2. Explain the characteristics features of Fabaceae family and write the uses of Groundnut

D3. Distinguishing characters of Pedaliaceae family and write the utilization of Gingelly

D4. Write the conomic importance of rice and cotton.

Chapter

20

Model Questions for Final Practical Examination

Final Practical Examination: BATCH-A

PBG 101 Introduction to Agricultural Botany

Time: 2.30 hrs **Max. Marks: 40**

1. Identify the specimen A to E and write the scientific name, family and economic part.($5\times1=5$)
2. Write short notes on ($5 \times 2 = 10$)
   (a) Phyllotaxy
   (b) Maize inflorescence
   (c) Androecium of Redgram and Soybean
   (d) Castor inflorescence
   (e) Extraction of sugar from Palmyra palm
3. Dissect the given specimen and write the botany of the specimen ($1 \times 10 = 10$)
4. Record (5 marks)
5. Assignment (5 marks)
6. Viva voce (5 marks)

Final Practical Examination: BATCH-B

PBG 101 Introduction to Agricultural Botany

Time: 2.30 hrs **Max. Marks: 40**

1. Identify the specimen A to E and write the scientific name, family and economic part
   ($5 \times 1 = 5$)
2. **Write short notes on** **($5 \times 2 = 10$)**
   (a) Venation
   (b) Sunflower inflorescence
   (c) Spadix of coconut
   (d) Cauliflower head
   (e) Androecium and Gynoecium Gossypium hirsutum

3. Dissect the given specimen and write the botany of the specimen: **(1 × 10 = 10)**

4. Record (5 marks)
5. Assignment (5 marks)
6. Viva voce (5 marks)

Final Practical Examination: BATCH-C

PBG 101 Introduction to Agricultural Botany

Time: 2.30 hrs **Max. Marks: 40**

1. Identify the specimen A to E and write the scientific name, family and economic part. (5 × 1 = 5)
2. **Write short notes on** **(5 × 2 = 10)**
   (a) Special features of Gingelly
   (b) Cymose inflorescence
   (c) Androecium and Gynoecium of Soybean
   (d) Cabbage leaf
   (e) Aestivation
3. Dissect the given specimen and write the botany of the specimen: **(1 × 10=10)**
4. Record (5 marks)
5. Assignment (5 marks)
6. Viva voce (5 marks)

Reference:

- Ashokkumar,. K,. Pandiyan, M and P. Sivakumar., 2019 Introduction to Agricultural Botany, Kalyani Publisher, New Delhi
- www. biologydiscussion.com
- Google search.com images